Music Recommendation and Discovery

Òscar Celma

Music Recommendation and Discovery

The Long Tail, Long Fail, and Long Play in the Digital Music Space

 Springer

Òscar Celma
BMAT
Bruniquer 49
08024 Barcelona
Spain
ocelma@bmat.com

ISBN 978-3-642-43953-7   ISBN 978-3-642-13287-2 (eBook)
DOI 10.1007/978-3-642-13287-2
Springer Heidelberg Dordrecht London New York

ACM Computing Classification (1998): H.3.3, G.2.2, H.5.5, I.7.2

Cover design: Claudia Lomelí Buyoli

Top Image: Tube Tags by Last.fm Limited. Main developers: Olivier Gillet, Hannah Donovan,
Norman Casagrande

Bottom Image: Last.fm similar artists graph by Dr. Tamás Nepusz (Department of Computer Science,
Royal Holloway, University of London)

Printed on acid-free paper

Springer is part of Springer Science+Business Media (www.springer.com)

Per l'Àlex

Foreword

In the last 15 years we have seen a major transformation in the world of music. Musicians use inexpensive personal computers instead of expensive recording studios to record, mix and engineer music. Musicians use the Internet to distribute their music for free instead of spending large amounts of money creating CDs, hiring trucks and shipping them to hundreds of record stores. As the cost to create and distribute recorded music has dropped, the amount of available music has grown dramatically. Twenty years ago a typical record store would have music by less than ten thousand artists, while today online music stores have music catalogs by nearly a million artists.

While the amount of new music has grown, some of the traditional ways of finding music have diminished. Thirty years ago, the local radio DJ was a music tastemaker, finding new and interesting music for the local radio audience. Now radio shows are programmed by large corporations that create playlists drawn from a limited pool of tracks. Similarly, record stores have been replaced by big box retailers that have ever-shrinking music departments. In the past, you could always ask the owner of the record store for music recommendations. You would learn what was new, what was good and what was selling. Now, however, you can no longer expect that the teenager behind the cash register will be an expert in new music, or even be someone who listens to music at all.

With so much more music available, listeners are increasingly relying on tools such as automatic music recommenders to help them find music. Instead of relying on DJs, record store clerks or their friends to get music recommendations, listeners are also turning to machines to guide them to new music. This raises a number of questions: How well do these recommenders work? Do they generate novel, interesting and relevant music recommendations? How far into the Long Tail do they reach? Do they create feedback loops that drive listeners to a diminishing pool of popular artists? What affect will automatic music recommenders have on the collective music taste?

In this book, Dr. Celma guides us through the world of automatic music recommendation. He describes how music recommenders work, explores some of the limitations seen in current recommenders, offers techniques for evaluating the effectiveness of music recommendations and demonstrates how to build effective recommenders by offering two real-world recommender examples. As we rely more and more on automatic music recommendation it is important for us to understand what makes a good music recommender and how a recommender can affect the world of music. With this knowledge we can build systems that offer novel, relevant and interesting music recommendations drawn from the entire world of available music.

Austin, TX, March 2010   *Paul Lamere*
Director of Developer Community
The Echo Nest

Preface

I met Timothy John Taylor (aka *Tyla*[1]) in 2000, when he established in Barcelona. He was playing some acoustic gigs, and back then I used to record a lot of concerts with a portable DAT. After a remarkable night, I sent him an email telling that I recorded the concert, so I could give him a copy. After all, we were living in the same city. He said "yeah sure, come to my house, and give me the CD's". So there I am, another nervous fan, trying to look cool while walking to his home...

My big brother, the first "music recommender" that I reckon, bought a vynil of *The Dogs d'Amour* in 1989. He liked the art cover—painted by the singer, *Tyla*—so he purchased it. The English rock band was just starting to be somewhat worldwide famous. They were in the UK charts, and also had played in the *Top of the Pops*. Then, they moved to L.A. to record an album. Rock magazines used to talk about their chaotic and unpredictable concerts, as well as the excesses of the members. Both my brother and myself felt in love with the band after listening to the album.

Tyla welcomes me at his home. We have a long chat surrounded by vintage guitars and amps, and unfinished paintings. I give him a few CDs including his last concert in Barcelona, as well as two other gigs that I recorded one year before. All of a sudden, he mentions the last project he is involved in: he has just re-joined the classic *Dogs d'Amour* line-up, after more than six years of inactivity. They were recording a new album. He was very excited and happy (ever after) about the project. I asked why they decided to re-join after all these years. He said: *We've just noticed how much interest there is on the Internet about the band.* Indeed, not being able to find the old releases made lot of profit for *eBayers* and the like.

When I joined *The Dogs d'Amour* Yahoo! mailing list in 1998 we were just a few dozens of fans that were discussing about the disbanded band, their solo projects, and related artists to fall upon. One day, the members of the band joined the list, too. It was like a big—virtual—family. Being part of the mailing list allowed us to have updated information about what the band was up to, and chat with them. One day they officially announced that the band was active again, and they had a new album

---

[1] http://www.myspace.com/tylaandthedogsdamour

ready (...I already knew that!). Sadly, the reunion only lasted for a couple of years, ending with a remarkable UK *Monsters of Rock* tour supporting *Alice Cooper*.

During the last few years, *Tyla* has released a set of solo albums. He has made his life based on viral marketing—including the help from fans—setting gigs, selling albums and paintings online, as well as in the concerts. Nowadays, he has much more control of the whole creative process than ever. The income allows him not needing any record label—he had some bad experiences with record labels back in the 80's epoch, when they controlled everything. Moreover, from the fan's point of view, living in the same city allowed me to help him in the creation process of a few albums. I even played some guitar bits in a couple of songs (and since then, I own one of his vintage Strat).

Up to now, he is still very active; he plays, paints, manages his tours, and a long etcetera. Yet, he is in the "long tail" of popularity. It is difficult to discover these type of artists when using music recommenders that do not support "less-known" artists. Indeed, for a music lover is very rewarding to discover *unknown* artists that fit into her music taste. In my case, music serendipity dates from 1989; with a cool album cover, and the good music taste of my brother. Now, I am willing to experience these feelings again...

Mexico City, March 2010   *Òscar Celma*
                                                      Chief Innovation Officer
                      Barcelona Music and Audio Technologies (BMAT)

Acknowledgements

This book wouldn't exist if it weren't for the the help and assistance of many people. At the risk of unfair omission, I want to express my gratitude to them. I would like to thank Ralf Gerstner, Senior Editor at Springer, for his perseverance and patience. Since 2007, Ralf has been interested in this work. He has been intermittently asking me about the status of the book since then. Well, here it is at last, Ralf.

This book would be much more difficult to read—except for the "Spanglish" experts—if it weren't for the excellent work of the following people: Paul Lamere, Owen Meyers, Terry Jones, Kurt Jacobson, Douglas Turnbull, Tom Slee, Kalevi Kilkki, Perfecto Herrera, Alberto Lumbreras, Daniel McEnnis, Xavier Amatriain, and Neil Lathia. They not only have helped me to improve the text, but have provided feedback, comments, suggestions, and—of course—criticism.

I would like to thank my colleagues from the Music Technology Group, where I spent ten years of my life working and doing research. Special thanks goes to Perfecto Herrera, Mohamed Sordo and Pedro Cano. They have provided me countless suggestions, and devoting much time to me during this long journey. Many thanks also to my BMAT colleagues, where I'm lucky enough to put into the real world the research I carried out while doing the PhD. Every day I feel I'm very fortunate to work with these talented people.

Last but not least, this work would have never been possible without the encouragement of my wife Claudia, who has provided me love and patience, and my lovely son Àlex (aka Alejandro, Ale, Cano or Cheto)—who altered my *last.fm* and *youtube* accounts with his favourite music. Nowadays, *Cri–Cri*, *Elmo* and *Barney*, coexists with *The Dogs d'Amour*, *Backyard Babies*, and other rock bands. I reckon that the two systems are a bit lost when trying to recommend me music and videos! Also, a special warm thanks to my parents Tere and Toni, my brother Marc, and the whole family in Barcelona and Mexico City.

Contents

Chapter 1
Introduction

1.1 Motivation

In recent years typical music consumption behaviour has changed dramatically. Personal music collections have grown, aided by technological improvements in networks, storage, portability of devices and Internet services. The number and the availability of songs have de-emphasised their value; it is usually the case that users own many digital music files that they have only listened to once, or not at all. It seems reasonable to suppose that with efficient ways to create a personalised order of users' collections, as well as ways to explore hidden "treasures" inside them, the value of their music collections would drastically increase.

Users own huge music collections that need proper storage and labelling. Search within digital collections gives rise to new methods for accessing and retrieving data. But, sometimes, there is no metadata—or only file names—to inform us of the audio content, and that is not enough for an effective navigation and discovery of the music collection. Users can, then, get lost searching in their own digital collections. Furthermore, the web is increasingly becoming the primary source of music titles in digital form. With millions of tracks available from thousands of websites, finding the *right* songs, and being informed of new music releases has become problematic.

On the digital music distribution front, there is a need to find ways of improving music retrieval and personalisation. Artist, title, and genre information might not be the only criteria to help music consumers find music they like. This is achieved using cultural or editorial metadata ("this artist is somehow related to that one"), or exploiting existing purchasing behaviour data ("since you bought this artist, you might also enjoy this one"). A largely unexplored—and potentially interesting—complement is using semantic descriptors automatically extracted from music files, or gathered from the community of users, via social tagging. All this information can be combined and used for music recommendation.

Ò. Celma, *Music Recommendation and Discovery*,
DOI 10.1007/978-3-642-13287-2_1, © Springer-Verlag Berlin Heidelberg 2010

1.1.1 Academia

With one early exception, Shardanand's masters thesis [1] published in 1994, research in music recommendation did not really begin until 2001. To show the increasing interest in this field, Table 1.1 presents the number of papers related to music recommendation since 2001. The table shows the list of related papers indexed by *Google Scholar*.[1] From 2004 onwards we have seen a sharp increase in the number of papers published in this field.

Year	Num. papers
1994	1
–	–
2001	3
2002	4
2003	3
2004	8
2005	14
2006	19
2007	21
2008	19
2009	19

Table 1.1 Number of scientific articles related to music recommendation, indexed by Google Scholar.

A closer look, focusing on the Music Information Retrieval (MIR) community, also shows an increasing interest in music recommendation and discovery. Table 1.2 shows the list of related papers, presented in ISMIR (International Society for Music Information Retrieval)[2] conferences since 2000. The early papers focused on content–based methods [2, 3], and user profiling aspects [4, 5]. Since 2005, research community attention has broadened to other areas, including: prototype systems [6–8], playlist generation [9–17], social tagging [18, 19], visual interfaces [20], music similarity networks [21–24], hybrid recommendation approaches [25–29], and sociological aspects [30–34]. The "Music Recommendation Tutorial" [35], presented in the ISMIR 2007 conference, summarises part of the work done in this field till then.

---

[1] We count, for each year, the number of results from http://scholar.google.com that contain "music recommendation" or "music recommender" in the title of the article. Accessed on Feburary 3rd, 2010

[2] http://www.ismir.net/

Year	Papers	References
2000	1	[4]
2001	0	–
2002	3	[2, 5, 9]
2003	0	–
2004	1	[3]
2005	4	[6, 7, 10, 12]
2006	6	[11, 13, 26, 30, 36, 37]
2007	7	[8, 20, 21, 25, 27, 31, 35]
2008	7	[14, 15, 18, 19, 22, 28, 32]
2009	7	[16, 17, 23, 24, 29, 33, 34]

Table 1.2 Papers related to music recommendation presented in the ISMIR conference since 2000.

1.1.2 Industry

Recommender systems play an important role in e-Commerce. Examples such as *Last.fm*, *Amazon*, or *Netflix*, where the provided recommendations are critical to retain users, show that most of the product sales result from the recommendations. Greg Linden, who implemented the first recommendation engine for *Amazon*, states[3]:

> *(Amazon.com) recommendations generated a couple orders of magnitude more sales than just showing top sellers.*

Since October 2006, this field enjoyed an increase of interest thanks to the *Netflix* competition. The competition offered a prize of $1,000,000 to those that improve their movie recommendation system.[4] Also, the *Netflix* competition provided the largest open dataset, containing more than 100 million movie ratings from anonymous users. The research community was challenged in developing algorithms to improve the accuracy of the current *Netflix* recommendation system. After 3 years of research, in July 2009, both *BellKor's Pragmatic Chaos*[5] and *The Ensemble*[6] teams did beat the 10% threshold, in both cases by blending several approaches to improve the overall result of the predictions.

---

[3] http://glinden.blogspot.com/2007/05/google-news-personalization-paper.html

[4] The goal was to reduce by 10% the Root mean squared error (RMSE) of the predicted movie ratings

[5] http://www2.research.att.com/~volinsky/netflix/bpc.html

[6] http://www.the-ensemble.com

1.1.2.1 State of the Music Industry

The *Long Tail*[7] is composed by a small number of popular items (the *hits*), and the rest are located in the tail of the curve [38]. The main goal of the Long Tail economics—originated by the huge shift from physical media to digital media, and the fall in production costs—is to make everything available, in contrast to the limitations of the *brick–and–mortar* stores. Thus, personalised recommendations and filters are needed to help users find the right content in the digital space.

On the music side, the 2007 "State of the Industry" report by Nielsen SoundScan presents some interesting information about music consumption in the United States [39]. Around 80,000 albums were released in 2007 (not counting music available in *Myspace.com*, and similar sites). However, traditional CD sales are down 31% since 2004—but digital music sales are up 490%. Indeed, 844 million digital tracks were sold in 2007, but only 1% of all digital tracks accounted for 80% of all track sales. Also, 1,000 albums accounted for 50% of all album sales, and 450,344 of the 570,000 albums sold were purchased less than 100 times.

Music consumption based on sales is biased towards a few popular artists. Ideally, by providing personalised filters and discovery tools to users, music consumption would diversify. There is a need to assist people to discover, recommend, personalise and filter the huge amount of music content. In this sense, *Echo Nest*,[8] and *BMAT*[9] companies, created out of prominent MIR research groups, provide specific solutions to solve these limitations.

1.2 What's the Problem with Music Recommendation?

Nowadays, we have an overwhelming number of choices of which music to listen to. We see this each time we browse a non-personalised music catalog, such as *Myspace* or *iTunes*. In *The Paradox of Choice* [40], Schwartz states that we, as consumers, often become paralyzed and doubtful when facing the overwhelming number of choices. There is a need to eliminate some of the choices, and this can be achieved by providing personalised filters and recommendations to ease users' decision.

---

[7] From now on, considered as a proper noun with capitalised letters

[8] http://echonest.com/

[9] http://bmat.com/

1.2.1 Music ≠ Movies and Books

Several music recommendation paradigms have been proposed in recent years, and many commercial systems have appeared with more or less success. Most of these approaches apply or adapt existing recommendation algorithms, such as collaborative filtering, into the music domain.

However, music is somewhat different from other entertainment domains, such as movies or books. Tracking users' preferences is mostly done implicitly, via their listening habits (instead of asking users to explicitly rate the items). Any user can consume an item (e.g., a track or a playlist) several times, even repeatedly and continuously. Regarding the evaluation process, music recommendation allows users instant feedback via brief audio excerpts.

The context is another big difference between music and the other two domains. People consume different music in different contexts; e.g. hard-rock early in the morning, classical piano sonatas while working, and Lester Young's cool jazz while having dinner. A music recommender has to deal with complex contextual information.

1.2.2 Predictive Accuracy vs. Perceived Quality

Current music recommendation algorithms try to accurately predict what people will want to listen to. However, these algorithms tend to recommend popular (or well-known to the user) artists, which decreases the user's perceived quality of the recommendations. The algorithms focus, then, on *predicting the accuracy* of the recommendations. That is, try to make accurate predictions about what a user could listen to, or buy next, independently of how useful the provided recommendations are to the user.

Figure 1.1 depicts this phenomenon. It shows *Amazon* similar albums for the Beatles' *White Album*,[10] based on the consumption habits of users. Top-30 recommendations for the Beatles' *White Album* are strictly made of other Beatles' albums (then suddenly, on the fourth page of results, there is the first non-Beatles album; *Exile on Main St.* by The Rolling Stones). For the system these are the most accurate recommendations and, ideally, the ones that maximise their goal—to make a user to buy more goods. Still, one might argue about the usefulness of the provided recommendations. In fact, the goals of a recommender are not always aligned with the goals of a listener. The goal of the *Amazon* recommender is to sell goods, whereas the goal for a user visiting *Amazon* may be to find some new and interesting music.

---

[10] http://www.amazon.com/Beatles-White-Album/dp/B000002UAX, accessed on October, 9th, 2008

Fig. 1.1 *Amazon* recommendations for The Beatles' "White Album".

1.3 Our Proposal

In this book we emphasise the user's *perceived quality*, rather than the system's *predictive accuracy* when providing recommendations. To allow users to discover new music, recommender systems should exploit the long tail of popularity (e.g., number of total plays, or album sales) that exists in any large music collection.

Figure 1.2 depicts the long tail of popularity, and how recommender systems should help us in finding interesting information [38]. Personalised filters assist us in filtering the available content, and in selecting those—potentially—novel and interesting items according to the user's profile. In this sense, the algorithm strengthens the user's *perceived quality* and usefulness of the recommendations. Two key elements to drive the users from the head to the tail of the curve are novelty, and personalised relevance. Effective recommendation systems should promote novel and relevant material (non-obvious recommendations), taken primarily from the tail of a distribution, rather than focus on accuracy.

1.3.1 Novelty and Relevance

Novelty is a property of a recommender system that promotes unknown items to a user. Novelty is the opposite of the user's *familiarity* with the recommended items.

Fig. 1.2 The Long Tail of items in a recommender system. An important role of a recommender is to drive the user from the head region (popular items) to the long tail of the *curve* [38].

Yet, serendipity, that is novel and relevant recommendations for a given user, cannot be achieved without taking into account the user profile. Personalised relevance filters the available content, and selects those (potentially novel) items according to user preferences.

Ideally, a user should also be familiar with some of the recommended items, to improve the confidence and trust in the system. The system should also give an explanation of why the items were recommended, providing higher confidence and transparency of novel recommendations. The difficult job for a recommender is, then, to find the proper level of familiarity, novelty and relevance for *each* user. This way, recommendations can use the long tail of popularity. Furthermore, the proper levels of familiarity, novelty and relevance for a user will change over time. As a user becomes comfortable with the recommendations, the amount of familiar items could be reduced.

1.3.2 Key Elements

Figure 1.3 depicts the main elements involved in our music recommendation approach. The item (or user) similarity graph defines the relationship among the items (or users). This information is used for recommending items (or like-minded people) to a given user, based on her preferences. The long tail curve models the popularity of the items in the dataset, according to the shared knowledge of the whole community. The user profile is represented along the popularity curve, using her list of preferred items.

Using the information from the similarity graph, the long tail of item popularity, and the user profile, we should be able to provide the proper level of familiarity, novelty and relevant recommendations to the users. Finally, an assessment of the provided recommendations is needed. This is done in two complementary ways. First, using a novel user-agnostic evaluation method based on the analysis of the

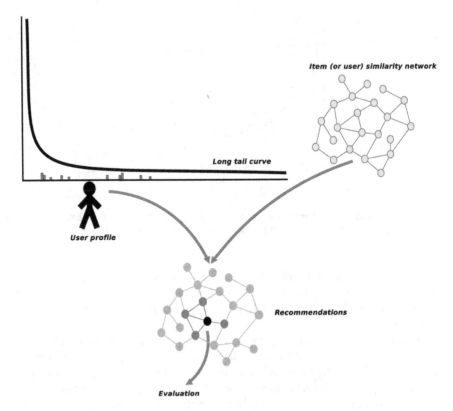

Fig. 1.3 Diagram that depicts the key elements of the book. It consists of the similarity graph, the long tail of item popularity, the user profile, the provided recommendations, and the evaluation part.

item (or user) similarity network, and the item popularity. Secondly, with a user-based evaluation, that provides feedback on the list of recommended items.

1.4 Summary of Contributions

The main contributions of this book are:

1. A *formal definition of the recommendation problem*, including the key elements that affect the recommendations provided by a system. Also, we present the existing recommendation methods to recommend items (and also like-minded people) to users, and we mention the pros and cons of each approach.
2. An instantiation of the general recommenddation problem for the *music domain*, highlighting its common use cases, as well as presenting different approaches to

user profiling and modelling, and how to link it with the the musical items (i.e. artists and songs). We also present the main elements that describe the musical items, using editorial, cultural, and audio-based descriptors.

3. A novel *network-based evaluation method* (or user-agnostic) for recommender systems, based on the analysis of the item (or user) similarity network, and the item popularity. This method has the following properties:

   a. it measures the novelty component of a recommendation algorithm,
   b. it makes use of complex network analysis to analyse the similarity graph,
   c. it models the item popularity curve,
   d. it combines both the complex network and the item popularity analysis to determine the underlying characteristics of the recommendation algorithm, and
   e. it does not require any user intervention in the evaluation process.

   We apply this evaluation method in the music domain, using large-scale artist and user similarity networks.

4. A *user-centric evaluation* based on the immediate feedback of the provided recommendations. This evaluation method has the following advantages (compared to other system-oriented evaluations):

   a. it measures the novelty factor of a recommendation algorithm in terms of user knowledge,
   b. it measures the relevance (e.g., like it or not) of the recommendations, and
   c. the users provide immediate feedback to the evaluation system, so the system can react accordingly.

   This method complements the previous, user-agnostic, evaluation approach. We use this method to evaluate three different music recommendation approaches (social-based, content-based, and a hybrid approach using expert human knowledge). In this experiment, 288 subjects rated their personalised recommendations in terms of novelty (*does the user know the recommended song/artist?*), and relevance (*does the user like the recommended song?*).

5. A music search engine, named *Searchsounds*, that allows users to discover unknown music mentioned on music-related blogs. *Searchsounds* provides keyword based search, as well as the exploration of similar songs using audio similarity.

6. A system prototype, named *FOAFing the music*, to provide music recommendations based on the user preferences and her listening habits. The main goal of the *Foafing the Music* system is to recommend, to discover and to explore music content; based on user profiling, context-based information (extracted from music related RSS feeds), and content-based descriptions (automatically extracted from the audio itself). *Foafing the Music* allows users to:

   a. get new music releases from *iTunes, Amazon, Yahoo Shopping*, etc.
   b. download (or stream) audio from MP3-blogs and Podcast sessions,
   c. discover music with *radio–a–la–carte* (i.e., personalised playlists),

d. view upcoming concerts happening near the user's location, and
e. read album reviews.

1.5 Book Outline

This book is structured as follows: Chap. 2 introduces the basics of the recommendation problem, and presents the general framework that includes user preferences and representation. Then, Chap. 3 adapts the recommendation problem to the music domain, and presents related work in this area. Once the users, items, and recommendation methods are presented, Chap. 4 introduces the Long Tail model and its usage in recommender systems. Chapters 5, 6 and 7 present the different ways of evaluating and comparing different recommendation algorithms. Chapter 5 presents the existing metrics for system-, network-, and user-centric approaches.

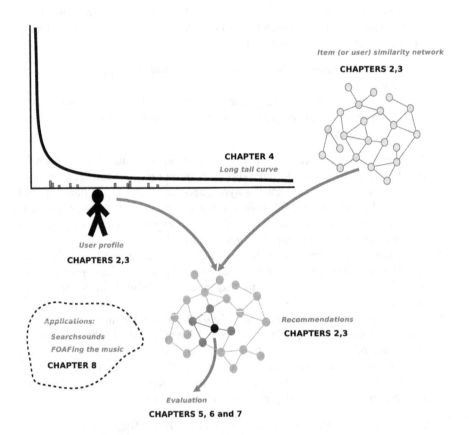

Fig. 1.4 Extension of Fig. 1.3 adding the corresponding chapters.

Then, Chap. 6 presents a complement to the classic system-centric evaluation, focusing on the analysis of the item (or user) similarity network, and its relationships with the popularity of the items. Chapter 7 complements the previous approach by entering the users in the evaluation loop, allowing them to evaluate the quality of the recommendations via immediate feedback. Chapter 8 presents two real prototypes. These systems, named *Searchsounds* and *FOAFing the music* show how to exploit music related content that is available on the web, for music discovery and recommendation. Chapter 9 draws some conclusions and discusses open issues and future work. To summarise the outline of the book, Fig. 1.4 presents an extension of Fig. 1.3, including the main elements of the book and its related chapters.

References

1. U. Shardanand, "Social information filtering for music recommendation," Master's thesis, Massachussets Institute of Technology, Cambridge, MA, September 1994.
2. B. Logan, "Content-based playlist generation: Exploratory experiments," in *Proceedings of 3rd International Conference on Music Information Retrieval*, (Paris, France), 2002.
3. B. Logan, "Music recommendation from song sets," in *Proceedings of 5th International Conference on Music Information Retrieval*, (Barcelona, Spain), 2004.
4. W. Chai and B. Vercoe, "Using user models in music information retrieval systems," in *Proceedings of 1st International Conference on Music Information Retrieval*, (Plymouth, MA, USA), 2000.
5. A. Uitdenbogerd and R. van Schnydel, "A review of factors affecting music recommender success," in *Proceedings of 3rd International Conference on Music Information Retrieval*, (Paris, France), 2002.
6. O. Celma, M. Ramirez, and P. Herrera, "Foafing the music: A music recommendation system based on RSS feeds and user preferences," in *Proceedings of 6th International Conference on Music Information Retrieval*, (London, UK), 2005.
7. R. van Gulik and F. Vignoli, "Visual playlist generation on the artist map," in *Proceedings of 6th International Conference on Music Information Retrieval*, (London, UK), pp. 520–523, 2005.
8. E. Pampalk and M. Goto, "Musicsun: A new approach to artist recommendation," in *Proceedings of 8th International Conference on Music Information Retrieval*, (Vienna, Austria), 2007.
9. S. Pauws and B. Eggen, "Pats: Realization and user evaluation of an automatic playlist generator," in *Proceedings of 3rd International Conference on Music Information Retrieval*, (Paris, France), 2002.
10. E. Pampalk, T. Pohle, and G. Widmer, "Dynamic playlist generation based on skipping behavior," in *Proceedings of 6th International Conference on Music Information Retrieval*, (London, UK), 2005.
11. E. Pampalk and M. Gasser, "An implementation of a simple playlist generator based on audio similarity measures and user feedback.," in *Proceedings of 7th International Conference on Music Information Retrieval*, (Victoria, Canada), pp. 389–390, 2006.
12. S. Pauws and S. van de Wijdeven, "User evaluation of a new interactive playlist generation concept," in *Proceedings of 6th International Conference on Music Information Retrieval*, (London, UK), pp. 638–643, 2005.
13. N. Oliver and L. Kregor-Stickles, "Papa: Physiology and purpose-aware automatic playlist generation," in *Proceedings of 7th International Conference on Music Information Retrieval*, (Victoria, Canada), pp. 250–253, 2006.

14. M. Niitsuma, H. Takaesu, H. Demachi, M. Oono, and H. Saito, "Development of an automatic music selection system based on Runner's step frequency," in *Proceedings of the 9th Conference on Music Information Retrieval*, (Philadelphia, Pennsylvania in USA), pp. 193–198, 2008.

15. M. G. Arthur Flexer, D. Schnitzer, and G. Widmer, "Playlist generation using start and end songs," in *Proceedings of the 9th Conference on Music Information Retrieval*, (Philadelphia, Pennsylvania in USA), pp. 219–224, 2008.

16. G. D. Francois Maillet, D. Eck and P. Lamere, "Steerable playlist generation by learning song similarity from radio station playlists," in *Proceedings of the 10th Conference on Music Information Retrieval*, (Kobe, Japan), pp. 345–350, 2009.

17. K. Bosteels, E. Pampalk, and E. E. Kerre, "Evaluating and analysing dynamic playlist generation heuristics using radio logs and fuzzy set theory," in *Proceedings of the 10th Conference on Music Information Retrieval*, (Kobe, Japan), pp. 351–356, 2009.

18. C. Baccigalupo, J. Donaldson, and E. Plaza, "Uncovering affinity of artists to multiple genres from social behaviour data," in *Proceedings of the 9th Conference on Music Information Retrieval*, (Philadelphia, Pennsylvania in USA), pp. 275–280, 2008.

19. P. Symeonidis, M. Ruxanda, A. Nanopoulos, and Y. Manolopoulos, "Ternary Semantic Analysis of Social Tags for Personalized Music Recommendation," in *Proceedings of the 9th Conference on Music Information Retrieval*, (Philadelphia, PA), pp. 219–224, 2008.

20. J. Donaldson, "Music recommendation mapping and interface based on structural network entropy.," in *Proceedings of 8th International Conference on Music Information Retrieval*, (Vienna, Austria), pp. 811–817, 2007.

21. A. Anglade, M. Tiemann, and F. Vignoli, "Virtual communities for creating shared music channels," in *Proceedings of 8th International Conference on Music Information Retrieval*, (Vienna, Austria), 2007.

22. B. Fields, K. Jacobson, C. Rhodes, and M. Casey, "Social playlists and bottleneck measurements: Exploiting musician social graphs using content-based dissimilarity and pairwise maximum flow values," in *Proceedings of the 9th Conference on Music Information Retrieval*, (Philadelphia, Pennsylvania in USA), pp. 559–564, 2008.

23. D. S. Klaus Seyerlehner, P. Knees and G. Widmer, "Browsing music recommendation networks," in *Proceedings of the 10th Conference on Music Information Retrieval*, (Kobe, Japan), pp. 129–134, 2009.

24. B. McFee and G. Lanckriet, "Heterogeneous embedding for subjective artist similarity," in *Proceedings of the 10th Conference on Music Information Retrieval*, (Kobe, Japan), pp. 513–518, 2009.

25. M. Tiemann and S. Pauws, "Towards ensemble learning for hybrid music recommendation," in *Proceedings of 8th International Conference on Music Information Retrieval*, (Vienna, Austria), 2007.

26. K. Yoshii, M. Goto, K. Komatani, T. Ogata, and H. G. Okuno, "Hybrid collaborative and content-based music recommendation using probabilistic model with latent user preferences," in *Proceedings of 7th International Conference on Music Information Retrieval*, (Victoria, Canada), pp. 296–301, 2006.

27. K. Yoshii, M. Goto, K. Komatani, T. Ogata, and H. G. Okuno, "Improving efficiency and scalability of model-based music recommender system based on incremental training," in *Proceedings of 8th International Conference on Music Information Retrieval*, (Vienna, Austria), 2007.

28. T. Magno and C. Sable, "A comparison of signal-based music recommendation to genre labels, collaborative filtering, musicological analysis, human recommendation, and random baseline," in *Proceedings of the 9th Conference on Music Information Retrieval*, (Barcelona, Spain), pp. 161–166, 2008.

29. K. Yoshii and M. Goto, " Continuous PLSI and smoothing techniques for hybrid music recommendation," in *Proceedings of the 10th Conference on Music Information Retrieval*, (Kobe, Japan), pp. 339–344, 2009.

30. S. J. Cunningham, D. Bainbridge, and A. Falconer, "More of an Art than a Science: Supporting the creation of playlists and mixes," in *Proceedings of 7th International Conference on Music Information Retrieval*, (Victoria, Canada), pp. 240–245, 2006.

31. D. McEnnis and S. J. Cunningham, "Sociology and music recommendation systems," in *Proceedings of 8th International Conference on Music Information Retrieval*, (Vienna, Austria), 2007.

32. P. Chordia, M. Godfrey, and A. Rae, "Extending content-based recommendation: The case of Indian classical music," in *Proceedings of the 9th Conference on Music Information Retrieval*, (Philadelphia, Pennsylvania in USA), pp. 571–576, 2008.

33. S. J. Cunningham and D. M. Nichols, "Exploring social music behavior: An investigation of music selection at parties," in *Proceedings of the 10th Conference on Music Information Retrieval*, (Kobe, Japan), pp. 747–752, 2009.

34. L. Barrington, R. Oda, and G. Lanckriet, "Smarter than genius? Human evaluation of music recommender systems," in *Proceedings of the 10th Conference on Music Information Retrieval*, (Kobe, Japan), pp. 357–362, 2009.

35. O. Celma and P. Lamere, "Music recommendation tutorial," in *Proceedings of 8th International Conference on Music Information Retrieval*, (Vienna, Austria), 2007.

36. X. Hu, J. S. Downie, and A. F. Ehmann, "Exploiting recommended usage metadata: Exploratory analyses," in *Proceedings of 7th International Conference on Music Information Retrieval*, (Victoria, Canada), pp. 19–22, 2006.

37. S. Pauws, W. Verhaegh, and M. Vossen, "Fast generation of optimal music playlists using local search," in *Proceedings of 7th International Conference on Music Information Retrieval*, (Victoria, Canada), pp. 138–143, 2006.

38. C. Anderson, *The Long Tail. Why the future of business is selling less of more*. New York, NY: Hyperion, 2006.

39. N. Soundscan, "State of the industry," *Nielsen Soundscan Report. National Association of Recording Merchandisers*, 2007.

40. B. Schwartz, *The Paradox of Choice: Why More Is Less*. Harper Perennial, January 2005.

Chapter 2
The Recommendation Problem

Generally speaking, the reason people could be interested in using a recommender system is that they have so many items to choose from—in a limited period of time—that they cannot evaluate all the possible options. A recommender should be able to select and filter all this information to the user. Nowadays, the most successful recommender systems have been built for entertainment content domains, such as: movies, music, or books.

This chapter is structured as follows: Sec. 2.1 introduces a formal definition of the recommendation problem. After that, Sec. 2.2 presents some use cases to stress the possible usages of a recommender. Section 2.3 presents the general model of the recommendation problem. An important aspect of a recommender system is how to model the user preferences and how to represent a user profile. This is discussed in Sec. 2.4. After that, Sec. 2.5 presents the existing recommendation methods to recommend items (and also like-minded people) to users. Finally, Sec. 2.6 presents some key elements that affect the recommendation problem.

2.1 Formalisation of the Recommendation Problem

Intuitively, the recommendation problem can be split into two subproblems. The first one is a prediction problem, and is about the estimation of the items' likeliness for a given user. The second problem is to recommend a list of N items—assuming that the system can predict likeliness for yet unrated items. Actually, the most relevant problem is the estimation. Once the system can estimate items into a totally ordered set, the recommendation problem reduces to list the top-N items with the highest estimated value.

- The *prediction problem* can be formalised as follows [1]: Let $U = \{u_1, u_2, \ldots u_m\}$ be the set of all users, and let $I = \{i_1, i_2, \ldots i_n\}$ be the set of all possible items that can be recommended.

Each user u_i has a list of items I_{u_i}. This list represents the items that the user has expressed her interests. Note that $I_{u_i} \subseteq I$, and it is possible that I_{u_i} be empty,[1] $I_{u_i} = \emptyset$. Then, the function, P_{u_a,i_j} is the predicted likeliness of item i_j for the active user u_a, such as $i_j \notin I_{u_a}$.

- The *recommendation problem* is reduced to bringing a list of N items, $I_r \subset I$, that the user will like the most (i.e. the ones with higher P_{u_a,i_j} value). The recommended list should not contain items from the user's interests, i.e. $I_r \cap I_{u_i} = \emptyset$.

The space I of possible items can be very large. Similarly, the user space U, can also be enormous. In most recommender systems, the prediction function is usually represented by a rating. User ratings are triples $\langle u, i, r \rangle$ where r is the value assigned—explicit or implicitly—by the user u to a particular item i. Usually, this value is a real number (e.g. from 0 to 1), a value in a discrete range (e.g. from 1 to 5), or a binary variable (e.g. *like/dislike*).

There are many approaches to solve the recommendation problem. One widely used approach is when the system stores the interaction (implicit or explicit) between a user and the item set. The system can provide informed guesses based on the interaction that all the users have provided. This approximation is called *collaborative filtering*. Another approach is to collect information describing the items and then, based on the user preferences, the system is able to predict which items the user will like the most. This approach is generally known as *content-based filtering*, as it does not rely on other users' ratings but on the description of the items. *Context-based filtering* approach uses contextual information about the items to describe them. Another approach is *demographic filtering*, that stereotypes the kind of users that like a certain item. Finally, the *hybrid* approach combines some of the previous approaches. Section 2.5 presents all these approaches.

2.2 Use Cases

Herlocker et al. identify some common usages of a recommender system [2]:

- *Find good items.* The aim of this use case is to provide a ranked list of items, along with a prediction of how much the user would like each item. Ideally, a user would expect some novel items that are unknown to the user, as well as some familiar items, too.
- *Find all good items.* The difference of this use case from the previous one is with regard the coverage. In this case, the false positive rate should be lower, thus presenting items with a higher precision.
- *Recommend sequence.* This use case aims at bringing to the user an ordered sequence of items that is pleasing as a whole. A paradigmatic example is a music recommender's automatic playlist generation.

---

[1] Specially when the user creates an account to a recommender system.

- *Just browsing*. In this case, users find pleasant to browse into the system, even if they are not willing to purchase any item. Simply as an entertainment.
- *Find credible recommender*. Users do not automatically trust a recommender. Then, they "play around" with the system to see if the recommender does the job well. A user interacting with a music recommender will probably search for one of her favourite artists, and check the output results (e.g. similar artists, playlist generation, etc.)
- *Express self*. For some users is important to express their opinions. A recommender that offers a way to communicate and interact with other users (via forums, weblogs, etc.) allows the self-expression of users. Thus, other users can get more information—from tagging, reviewing or blogging processes—about the items being recommended to them.
- *Influence others*. This use case is the most negative of the ones presented. There are some situations where users might want to influence the community in viewing or purchasing a particular item. For example: Movie studios could rate high their latest new release, to push others to go and see the movie. In a similar way, record labels could try to promote their artists into the recommender.

All these use cases are important when evaluating a recommender. The first task of the evaluators should be to identify the most important use cases for which the recommender will be used, and base their decisions on that.

2.3 General Model

The main elements of a recommender are the *users* and the *items*. Users need to be modelled in a way that the recommender can exploit their profiles and preferences. Besides, an accurate description of the items is also crucial to achieve good results when recommending items to users.

Figure 2.1 describes the major entities and processes involved in the recommendation problem. The first step is to model both the users and the items, and it is presented in Sec. 2.4. After that, two type of recommendations can be computed; presenting the recommended items to the user (*Top-N predicted items*), and matching like-minded people (*Top-N predicted neighbours*). Once the user gets a list of recommended items, she can provide feedback, so the system can update her profile accordingly (profile adaptation).

2.4 User Profile Representation

There are two key elements when describing user preferences: the generation and maintenance of the profiles, and the exploitation of the profile using a recommendation algorithm [3]. On the one hand, profile generation involves the representation,

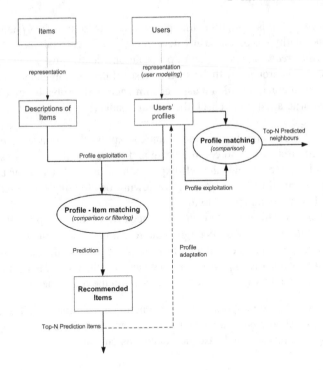

Fig. 2.1 General model of the recommendation problem.

initial generation, and adaptation techniques. On the other hand, profile exploitation involves the information filtering method used (i.e. the recommendation method), the matching between a user profile and the items, and the matching between user profiles (i.e. creation of neighbourhoods).

There are several approaches to represent user preferences. For instance, using the history of purchases in an e-Commerce website, web usage mining (analysis of the links, and time spent in a webpage), the listening habits (songs that a user listens to), etc.

2.4.1 Initial Generation

2.4.1.1 Empty

An important aspect of a user profile is its initialisation. The simplest way is to create an empty profile, that will be updated as soon as the user interacts with the system. However, the system will not be able to provide any recommendation until the user has been into the system for a while.

2.4.1.2 Manual

Another approach is to manually create a profile. In this case, a system might ask to the users to register their interests (via tags, keywords or topics) as well as some demographic information (e.g. age, marital status, gender, etc.), geographic data (city, country, etc.) and psychographic data (interests, lifestyle, etc.). The main drawback is the user's effort, and the fact that maybe some interests could still be unknown by the user himself.

2.4.1.3 Data Import

To avoid the manually creation of a profile, the system can ask to the user for available, external, information that already describes her. In this case, the system only has to import this information from the external sources that contain relevant information of the user.[2] Besides, there have been some attempts to allow users to share their own interests in a machine-readable format (e.g. XML), so any system can use it and extend it. An interesting proposal is the *Attention Profile Markup Language* (APML).[3]

The following example[4] shows a fragment of an APML file derived from the listening habits of a last.fm user.[5] The APML document contains a tag cloud representation created from the tags defined in the user's top artists.

```
<Profile name="music">
 <ImplicitData>
  <Concepts>
   <Concept key="rock" value="1.0" />
   <Concept key="hard_rock" value="0.41770712" />
   <Concept key="sleaze_rock" value="0.39724553" />
   <Concept key="rock_n_roll" value="0.3311153" />
   <Concept key="glam_rock" value="0.23445463" />
   <Concept key="classic_rock" value="0.2062444" />
   <Concept key="singer_songwriter" value="0.17533751" />
   <Concept key="alternative" value="0.1623969" />
   ...
  </Concepts>
 </ImplicitData>
</Profile>
```

Listing 2.1 Example of a user profile in APML.

---

[2] A de-facto standard, in the Semantic Web community, is the Friend of a Friend initiative (FOAF). FOAF provides conventions and a language "to tell" a machine the sort of things that a user says about herself. This approach is the one been used in our prototype, presented in Chap. 8

[3] http://www.apml.org

[4] Generated via http://TasteBroker.org, accessed on January, 10th 2008

[5] http://research.sun.com:8080/AttentionProfile/apml/last.fm/ocelma, accessed on January, 10th 2008

Fig. 2.2 Example of a pre-defined training set to model user preferences when a user created an account in *iLike*.

2.4.1.4 Training Set

Another method to gather information is using a pre-defined training set. The user has to provide feedback to concrete items, marking them as relevant or irrelevant to her interests. The main problem, though, is to select representative examples. For instance, in the music domain, the system might ask for concrete genres or styles, and filter a set of artists to be rated by the user. Figure 2.2 shows an example of the *iLike* music recommender. Once a user created an account, the system presents a list of artists that the user has to rate. This process is usually perceived by the users as a tedious and unnecessary work. Yet, it gives some information to the system to avoid the user cold-start problem (see Sec. 2.6 for more details).

2.4.1.5 Stereotyping

Finally, the system can gather initial information using *stereotyping*. This method resembles to a clustering problem. The main idea is to assign a new user into a cluster of similar users that are represented by their stereotype, according to some demographic, geographic, or psychographic information.

2.4.2 Maintenance

Once the profile has been created, it does not remain static. Therefore, user's interests might (and probably will) change. A recommender system needs up-to-date information to automatically update a user profile. User feedback may be explicit or implicit.

2.4.2.1 Explicit Feedback

One option is to ask to the users for relevance feedback about the provided recommendations. Explicit feedback usually comes in the form of ratings. This type of feedback can be positive or negative. Usually, users provide more positive feedback, although negative examples can be very useful for the system.

Ratings can be in a discrete scale (e.g. from 0 to N), or a binary value (*like/dislike*). Yet, it is proved that sometimes users rate inconsistently [4], thus ratings are usually biased towards some values, and this can also depend on the user perception of the ratings' scale. Inconsistency in the ratings arouse a natural variability when the system is predicting the ratings. Herlocker et al. presents a study showing that even the best algorithm could not get beyond a *Root mean squared error* (RMSE) of 0.73, on a five-point scale [2]. In [5] the authors present an experiment where users have to rate the items several times over a period of time. Then, they calculated the RMSE between different trials. RMSE ranged between 0.557 and 0.8156, depending on the ellapsed time (an improvement of 10% in the Netflix prize equals to an RMSE of 0.8563, so any algorithm has a very small margin of error). User consistency over time has strong consequences for recommender systems based on maximising the predictive accuracy, by trying to minimise the RMSE.

Another way to gather explicit feedback is to allow users to write comments and opinions about the items. In this case, the system can present the opinions to the target user, along with the recommendations. This extra piece of information eases the decision–making process of the target user, although she has to read and interpret other users' opinions.

2.4.2.2 Implicit Feedback

A recommender can also gather implicit feedback from the user. A system can infer the user preferences passively by monitoring user's actions. For instance, by analysing the history of purchases, the time spent on a webpage, the links followed by the user, the mouse movements, or analysing a media player usage (tracking the *play*, *pause*, *skip* and *stop* buttons).

However, negative feedback is not reliable when using implicit feedback, because the system can only observe positive (implicit) feedback, by analysing user's actions. On the other hand, implicit feedback is not as intrusive as explicit feedback.

2.4.3 Adaptation

The system has to adapt to the changes of the users' profiles. The techniques to adapt to new interests and forget the old ones can be done in three different ways. First, done manually by the user, although this requires some extra effort to the user. Secondly, by adding new information into the user profiles, while keeping the old interests. Finally, by gradually forgetting the old interests and promoting the new ones [6].

2.5 Recommendation Methods

Once the user profile is created, the next step is to exploit user preferences, to provide interesting recommendations. User profile exploitation is tightly related with the method for filtering information. The method adopted for information filtering has led to the standard classification of recommender systems, that is: demographic filtering, collaborative filtering, content-based and hybrid approaches. We add another method, named context-based, which recently has grown popularity due to the feasibility of gathering external information *about* the items (e.g. gathering information from weblogs, analysing the reviews about the items, etc.).

The following sections present the recommendation methods for *one* user. It is worth to mention that another type of (group-based) recommenders also exist. These recommenders focus on providing recommendations to a group of users, thus trying to maximise the overall satisfaction of the group [7, 8].

2.5.1 Demographic Filtering

Demographic filtering can be used to identify the kind of users that like a certain item [9]. For example, one might expect to learn the type of person that likes a certain singer (e.g. finding the stereotypical user that listens to *Jonas Brothers*[6] band). This technique classifies the user profiles in clusters according to some personal data (age, marital status, gender, etc.), geographic data (city, country) and psychographic data (interests, lifestyle, etc.). An early example of a demographic filtering system is the Grundy system [9]. Grundy recommended books based on personal information gathered from an interactive dialogue.

---

[6] http://www.jonasbrothers.com/

2.5.1.1 Limitations

The main problems of this method is that a system recommends the same items to people with similar demographic profiles, so recommendations are too general (or, at least, not very specific for a given user profile). Another drawback is the generation of the profile, that needs some effort from the user. Some approaches try to get (unstructured) information from the user's homepage, weblog, etc. In this case, text classification techniques are used to create the clusters and classify the users [10]. All in all, this is the simplest recommendation method.

2.5.2 Collaborative Filtering

Collaborative filtering (CF) approach predicts user preferences for items by learning past user-item relationships. That is, the user gives feedback to the system, so the system can provide informed guesses based on the feedback (e.g. ratings) that other users have provided.

The first system that implemented the collaborative filtering method was the *Tapestry* project at Xerox PARC [11]. The project coined the *collaborative filtering* term. Other early systems are: a music recommender named *Ringo* [12, 13], and Group Lens, a system for rating USENET articles [14]. A compilation of other relevant systems from that time period can be found in [15].

CF methods work by building a matrix M, with n items and m users, that contains the interaction (e.g. ratings, page views, plays, etc.) of the users with the items. Each row represents a user profile, whereas the columns are items. The value M_{u_a,i_j} is the rating of the user u_a for the item i_j. Figure 2.3 depicts the matrix of user-item ratings.

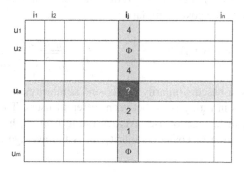

Fig. 2.3 User-item matrix for the collaborative filtering approach.

2.5.2.1 Item-Based Neighbourhood

Item-based method exploits the similarity among the items. This method looks into
the set of items that a user has rated, and computes the similarity among the target
item (to decide whether is worth to recommend it to the user or not). Figure 2.4
depicts the co-rated items from different users. In this case it shows the similarity
between items i_j and i_k. Note that only users u_2 and u_i are taken into account, but
u_{m-1} is not because it has not rated both items.

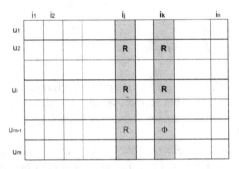

Fig. 2.4 User-item matrix with co-rated items for item-based similarity. To compute the similarity
between items i_j and i_k, only users u_2 and u_i are taken into account, but u_{m-1} is not because it has
not rated both items (i_k rating value is not set).

The first step is to obtain the similarity between two items, i and j. This similarity
can be calculated using cosine similarity, Pearson correlation, adjusted cosine, or
computing the conditional probability, $P(j|i)$. Let the set of users who rated i and
j be denoted by U, and $r_{u,i}$ denotes the rating of user u on item i. Equation (2.1)
shows the definition of the cosine similarity:

$$sim(i,j) = cos(\mathbf{i},\mathbf{j}) = \frac{\mathbf{i} \cdot \mathbf{j}}{\|i\| * \|j\|} = \frac{\sum_{u \in U} r_{u,i} r_{u,j}}{\sqrt{\sum_{u \in U} r_{u,i}^2} \sqrt{\sum_{u \in U} r_{u,j}^2}} \qquad (2.1)$$

However, for the item-based similarity, the cosine similarity does not take into account the differences in rating scale between different users. The adjusted cosine
similarity (Eq. 2.2) makes use of user average rating from each co-rated pair, and
copes with the limitation of cosine similarity. $\bar{r}_u$ is the average rating of the u-th
user:

$$sim(i,j) = \frac{\sum_{u \in U} (r_{u,i} - \bar{r}_u)(r_{u,j} - \bar{r}_u)}{\sqrt{\sum_{u \in U} (r_{u,i} - \bar{r}_u)^2} \sqrt{\sum_{u \in U} (r_{u,j} - \bar{r}_u)^2}} \qquad (2.2)$$

Correlation-based similarity commonly uses the Pearson correlation. The correlation between two variables reflects the degree to which the variables are related.

Equation (2.3) defines the correlation similarity. $\bar{r}_i$ is the average rating of the i-th item:

$$sim(i,j) = \frac{Cov(i,j)}{\sigma_i \sigma_j} = \frac{\sum_{u \in U}(r_{u,i} - \bar{r}_i)(r_{u,j} - \bar{r}_j)}{\sqrt{\sum_{u \in U}(r_{u,i} - \bar{r}_i)^2}\sqrt{\sum_{u \in U}(r_{u,j} - \bar{r}_j)^2}} \qquad (2.3)$$

Equation (2.4) defines similarity using conditional probability, $P(j \mid i)$:

$$sim(i,j) = P(j \mid i) \simeq \frac{f(i \cap j)}{f(i)} \qquad (2.4)$$

where $f(X)$ equals to the number of customers who have purchased the item set X. This is the only metric that is asymmetric. That is, $sim(i,j) \neq sim(j,i)$.

Once the similarity among the items has been computed, the next step is to predict to the target user, u, a value for the active item, i. A common way is to capture how the user rates the similar items of i. Let $S^k(i;u)$ denote the set of k neighbours of item i, that the user u has rated. The predicted value is based on the weighted sum of the user's ratings, $\forall j \in S^k(i;u)$. Equation (2.5) shows the predicted value for item i to user u.

$$\hat{r}_{u,i} = \frac{\sum_{j \in S^k(i;u)} sim(i,j) r_{u,j}}{\sum_{j \in S^k(i;u)} sim(i,j)} \qquad (2.5)$$

2.5.2.2 User-Based Neighbourhood

The predicted rating value of item i, for the active user u, $\hat{r}_{u,i}$, can also be computed by taking into account those users that are similar (are like-minded) to u. Equation (2.6) shows the predicted rating score of item i, for user u. $\bar{r}_u$ is the average rating of user u, and $r_{u,i}$ denotes the rating of the user u for the item i. Let $S^k(u)$ denote the set of k neighbours for user u, $\hat{r}_{u,i}$ is defined as:

$$\hat{r}_{u,i} = \bar{r}_u + \frac{\sum_{v \in S(u)^k} sim(u,v)(r_{v,i} - \bar{r}_v)}{\sum_{v \in S(u)^k} sim(u,v)} \qquad (2.6)$$

Yet, to predict $\hat{r}_{u,i}$, the algorithm needs to know beforehand the set of users similar to u, $S^k(u)$, as well as how similar they are, $sim(u,v)$.

The most common approaches to find the neighbours in either item-based $S^k(i;u)$, or user-based neighbourhood $S(u)^k$ approaches are Pearson correlation (Eq. 2.3), cosine similarity (Eq. 2.1), and matrix factorisation approaches.

2.5.2.3 Matrix Factorisation

Matrix factorisation techniques—such as Singular Value Decomposition (SVD), Non-negative Matrix Factorisation (NMF), or Principal Component Analysis (PCA)—are useful when the M user-item matrix is sparse, which is very common in any recommender system. Any factorisation technique aims at reducing the dimensionality of the original matrix, generating two matrices U and V that approximate the original matrix. For instance, Singular Value Decomposition (SVD) method computes matrices $(n \times k)U$ and $(m \times k)V$, for a given number of dimensions k, such that:

$$M = U\Sigma V^T, \tag{2.7}$$

where Σ is a diagonal matrix containing the singular values of M.

Matrix decomposition is not unique. There are different methods to approximate Eq. (2.7). For instance, Least squares method requires that the estimated matrices has to deviate as little as possible from M. Or stochastic gradient descent, that iteratively approximates matrices U and V, and updates them in order to minimise the squared error between the predictions and the actual ratings [16].

Once the matrix has been reduced to k dimensions, the predicted rating value of item i for a user u, $\hat{r}_{u,i}$, can be approximated as the dot product between the user's $U_u \in \mathbb{R}^k$ and the item's feature vector, $V_i \in \mathbb{R}^k$.

$$\hat{r}_{u,i} = U_u \cdot V_i^T = \sum_{f=0}^{k} U_{u,f} V_{f,i} \tag{2.8}$$

Matrix factorisation is used in collaborative filtering to deal with the sparsity problem, by reducing the matrix M to k dimensions (or latent factors). Furthermore, matrix factorisation can also be applied to derive user or item similarity in the reduced k-space, using cosine similarity in the U (users' latent factors) or V (items) matrices.

2.5.2.4 Limitations

Collaborative filtering is one of the most used recommendation methods, yet it presents some drawbacks:

- *Data sparsity* and *high dimensionality* are two inherent properties of the datasets. With a relative large number of users and items, the main problem is the low coverage of the users' ratings among the items. It is common to have a sparse user-item matrix of 1% (or less) coverage. Thus, sometimes it can be difficult to find reliable neighbours (specially for user-based CF).
- Another problem, related with the previous one, is that users with *atypical* tastes (that vary from the norm) will not have many users as neighbours. Thus, this will lead to poor recommendations. This problem is also known as *gray sheep* [17].

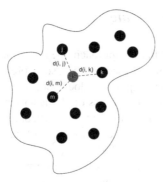

Fig. 2.5 Distance among items using content-based similarity.

- *Cold-start problem* This problem appears for both elements of a recommender: users and items. Due to CF is based on users' ratings, new users with only a few ratings become more difficult to categorise. The same problem occurs with new items, because they do not have any rating when added to the collection. These cannot be recommended until users start rating it. This problem is known as the *early-rater* problem [18]. Moreover, the first user that rates new items gets only little benefit (this new item does not match with any other item yet).
- CF is based only on the feedback provided by the users (in terms of ratings, purchases, downloads, etc.), and does not take into account the description of the items. It is a subjective method that aggregates the social behaviour of the users, thus commonly leading towards recommending the most popular items.
- Related with the previous issue, the *popularity bias* is another problem that commonly happens in CF. It is analogous to the *rich gets richer* paradigm. Popular items of the dataset are similar to (or related with) lots of items. Thus, it is more probable that the system recommends these popular items. This clearly happens for item-based similarity using conditional probability (defined in Eq. 2.4). The main drawback is that the recommendations are sometimes biased towards popular items, thus not exploring the Long Tail of unknown items. Sometimes, these less-popular items could be more interesting and novel for the users.
- Given the interactive behaviour of CF systems, previous social interaction influences the current user behaviour, which, in turn, feedbacks into the system, creating a loop. This issue is also known as *feedback loop* [19]. This effect has strong consequences when the system starts gathering initial feedback from the users. Indeed, the *early raters* have effects on the recommendations that the incoming users will receive when entering to the system.

2.5.3 Content-Based Filtering

In the content-based (CB) filtering approach, the recommender collects information describing the items and then, based on the user's preferences, it predicts which items the user could like. This approach does not rely on other user ratings but on the description of the items. The process of characterising the item data set can be automatic (e.g. extracting features by analysing the content), based on manual annotations by the domain experts. The key component of this approach is the similarity function among the items (see Fig. 2.5).

Initial CB approaches have its roots in the information retrieval (IR) field. The early systems focused on the text domain, and applied techniques from IR to extract *meaningful* information from the text. Yet, recently have appeared some solutions that cope with more complex domains, such as music. This has been possible, partly, because the multimedia community emphasised on and improved the feature extraction and machine learning algorithms.

The similarity function computes the distance between two items. Content-based similarity focus on an objective distance among the items, without introducing any subjective factor into the metric (as CF does). Most of the distance metrics deal with numeric attributes, or single feature vectors. Some common distances, given two feature vectors x and y, are: Euclidean (Eq. 2.9), Manhattan (Eq. 2.10), Chebychev (Eq. 2.11), cosine distance for vectors (see previously defined Eq. 2.1), and Mahalanobis distance (Eq. 2.12).

$$d(x,y) = \sqrt{\sum_{i=1}^{n} (x_i - y_i)^2} \tag{2.9}$$

$$d(x,y) = \sum_{i=1}^{n} |x_i - y_i| \tag{2.10}$$

$$d(x,y) = max_{i=1..n} |x_i - y_i| \tag{2.11}$$

$$d(x,y) = \sqrt{(x-y)^T S^{-1} (x-y)} \tag{2.12}$$

Euclidean, Manhattan and Chebychev distance are assuming that the attributes are orthogonal. The Mahalanobis distance is more robust to the dependencies among attributes, as it uses the covariance matrix S.

If the attributes are nominal (not numeric), a delta function can be used. A simple definition of a delta function could be: $\delta(a,b) = 0 \Leftrightarrow a = b$, and $\delta(a,b) = 1$ otherwise. Then, a distance metric among nominal attributes can be defined as:

$$d(x,y) = \omega \sum_{i=1}^{n} \delta(x_i, y_i), \tag{2.13}$$

where ω is a reduction factor, e.g. $\frac{1}{n}$).

Finally, if the distance to be computed has to cope with both numeric and nominal attributes, then the final distance has to combine two equations (2.13 for nominal attributes and one of 2.9...2.12 for numeric attributes). In some cases, items are not modelled with a single feature vector, but using a bag-of-vectors, a time series, or a probability distribution over the feature space.

2.5.3.1 Limitations

CB approach presents some drawbacks:

- The *cold-start* problem occurs when a new user enters to the system. The system has yet to adapt to the user preferences.
- The *gray-sheep* problem (users with atypical tastes) can occur, too, depending on the size of the collection, or if the collection is biased towards a concrete genre.
- Another potential caveat could be the *novelty* problem. Assuming that the similarity function works accurately, then one might assume that a user will always receive items *too* similar to the ones in her profile. To cope with this shortcoming, the recommender should use other factors to promote the *eclecticness* of the recommended items.
- Depending on the domain complexity, another drawback is the limitation of the features that can be (automatically) extracted from the objects. For instance in the multimedia arena, nowadays, is still difficult to extract high-level descriptors with a clear meaning for the user. Music analysis is not ready yet to accurately predict the *mood* of a song but, on the other hand, it does the job well when dealing with descriptors such as: harmony, rhythm, etc. Thus, even though an item description might not be meaningful for a user, still its description is useful to compute item similarity.
- Another shortcoming is that the recommender is focused on finding similarity among items, using only features describing the items. This means that subjectivity (or personal opinions) is not taken into account when the recommendations are computed.

CB methods solve some of the shortcomings of the collaborative filtering. The early-rater problem disappears. When adding a new item into the collection—and computing the similarity among the rest of the items—it can be recommended without being rated by any user. The popularity bias is solved too. Because there is no human intervention in the process, all the items are considered (in principle) to be of equal importance.

2.5.4 Context-Based Filtering

2.5.4.1 Context vs. Content

Context is any information that can be used to characterise the situation of an entity
[20]. Context-based recommendation uses, then, contextual information to describe
and characterise the items. To compare content and context-based filtering, one example is the different methods used for email spam detection. The common one is
based on the text analysis of the mail (i.e. content-based), whereas context filtering
does not deal with the content of the mail. It rather uses the context of the Simple
Mail Transfer Protocol (SMTP) connection to decide whether an email should be
marked as spam or not.

Now, we briefly outline two techniques, named Web mining and Social tagging,
that can be used to derive similarity among the items (or users). Web mining is based
on analysing the available content on the Web, as well as the usage and interaction
with the content. Social tagging mines the information gathered from a community
of users that tag items.

2.5.4.2 Web Mining

Web mining techniques aim at discovering interesting and useful information from
the analysis of the content and its usage. Kosala and Blockeel identify three different
web mining categories: content, structure and usage mining [21].

- *Web content mining* includes text, hypertext, markup, and multimedia mining.
  Some examples are: opinion extraction (sentiment analysis), weblog analysis,
  mining customer reviews, extract information from forums or chats, topic recognition and demographic identification (gender, age, etc.), and trend identification.
  Item similarity can be derived out of the analysis of this information.
- *Web structure mining* focuses on link analysis (in- and out- links). That is the network topology analysis (e.g. hubs, authorities), and the algorithms that exploits
  the topology (e.g. Hits and PageRank).
- *Web usage mining* uses the information available on session logs. This information can be used to derive user habits and preferences, link prediction, or item
  similarity based on co-occurrences in the session log. Thus, web usage mining
  can determine sequential patterns of usage (e.g. "people who visit this page also
  visited this one"). For instance, Mobasher et al. use association rules to determine the sequential patterns of web pages, and recommend web pages to users
  [22].

Combining these three approaches, a recommender system derives the similarity
among the items (e.g. items that co-occur in the same pages, items that are visited in
the same session log, etc.) and also models the users, based on their interaction with
the content. If the information about the content is in textual form, classic measures
from Information Retrieval can be applied to characterise the items. For instance,

vector space-based models can be used to model both the items and the user profile. Then, similarity between an item description (using the bag-of-words model) and a user profile can be computed using, for instance, cosine based similarity.

Cosine similarity between an item i_j, and a user profile u_i is defined as:

$$sim(u_i, i_j) = \frac{\sum_t w_{t,u_i} w_{t,i_j}}{\sqrt{\sum_t w_{t,u_i}^2} \sqrt{\sum_t w_{t,i_j}^2}} \qquad (2.14)$$

A common term weighting function, $w_{i,j}$, is the TF-IDF. TF stands for Term Frequency, whereas IDF is the Inverse Document Frequency [23]. The term frequency in a given document measures the importance of the term i within that particular document. Equation (2.15) defines TF:

$$TF = \frac{n_i}{\sum_k n_k} \qquad (2.15)$$

with n_i being the number of occurrences of the considered term, and the denominator is the number of occurrences of all the terms in the document.

The Inverse Document Frequency, IDF, measures the general importance of the term, in the whole collection of items:

$$IDF = \log \frac{|D|}{|(d_i \supset t_i)|} \qquad (2.16)$$

where $|D|$ is the total number of items, and the denominator counts the number of items where t_i appears. Finally, the weighting function $w_{t,j}$, of a term t in the item description d_j is computed as:

$$w_{t,j} = TF \cdot IDF \qquad (2.17)$$

Another useful measure to compute item similarity is the *Pointwise mutual information* (PMI). PMI estimates the semantic similarity between a pair of terms by how frequently they co-occur. The PMI of two terms i and j quantifies the discrepancy between their joint distribution probability, versus their individual distribution probability (assuming independence):

$$PMI(i, j) = \log \frac{p(i, j)}{p(i)p(j)} \qquad (2.18)$$

PMI measure is symmetric, that is $PMI(x,y) = PMI(y,x)$.

2.5.4.3 Social tagging

Social tagging (also known as Folksonomy, or Collaborative tagging) aims at annotating web content using tags. Tags are freely chosen keywords, not constrained to a predefined vocabulary. A bottom-up classification emerge when grouping all the annotations (tags) from the community of users; the wisdom of the crowds.

Recommender systems can derive social tagging data to derive item (or user) similarity.

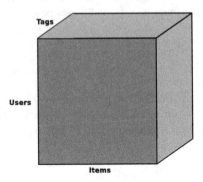

Fig. 2.6 The user-item-tag cube. A 3-order tensor containing $\langle user, item, tag \rangle$ triples.

When users tag items, we get tuples of $\langle user, item, tag \rangle$. These triples conform a 3-order matrix (also called *tensor*; a multidimensional matrix). Figure 2.6 depicts a 3-order tensor, containing the tags that users apply to items.

There are two main approaches to use social tagging information to compute item (and user) similarity. These are:

1. Unfold the 3-order tensor in three bidimensional matrices (user-tag, item-tag and user-item matrices), and
2. Directly use the 3-order tensor.

Unfolding the 3-order tensor consists on decomposing the multidimensional data into the following bidimensional matrices:

- *User-Tag* (*U* matrix). $U_{i,j}$ contains the number of times user i applied the tag j. Using matrix U, a recommender system can derive a user profile (e.g. a tag cloud for each user, denoting her interests, or the items she tags). U can also be used to compute user similarity, by comparing two user tag clouds of interests (using, for instance, cosine similarity between the two user vectors).
- *Item-Tag* (*I* matrix). $I_{i,j}$ contains the number of times an item i has been tagged with tag j. The matrix I contains the contextual description of the items, based on the tags that have been applied to. Matrix I can be used to compute item or user similarity. As an example, Fig. 2.7 shows a way to derive user similarity from I, using their *top-N* artists in *last.fm*. Figure 2.7 depicts two user tag clouds (top and middle images) and their intersection (bottom image), using matrix I. In this example, users' tag clouds are derived from the *last.fm* listening habits, using their top-N most listened artists—in this case, the items in I. The third image (bottom) shows the tags that co-occur the most in the two profiles. Similarity between the two users is done by constructing a new tag

Fig. 2.7 Two examples of users' tag clouds derived from their *last.fm* listening habits. *Top* and *middle* images show two *last.fm* user tag clouds. The third image (*bottom*) shows the tags that co-occur the most in the two profiles. According to Anthony Liekens' algorithm, the similarity value between *ocelma* and *lamere* last.fm users is 70.89%. Image courtesy of Anthony Liekens, taken from http://anthony.liekens.net/pub/scripts/last.fm/compare.php.

vector where each tag's weight is given by the minimum of the tag's weights in the user's vectors. Using this approach, the similarity value between *ocelma* and *lamere last.fm* users is 70.89%. Another similarity metric could be the cosine distance, using TFxIDF to weight each tag.

- *User-Item* (R binary matrix). $R_{i,j}$ denotes whether the user i has tagged the item j. In this case, classic collaborative filtering techniques can be applied on top of R.

To recap, item similarity using matrix I, or user similarity derived from U or I, can be computed using cosine-based distance (see Eq. 2.1), or also by applying dimensionality reduction techniques—to deal with the sparsity problem—such as Singular Value Decomposition (SVD), or Non-negative matrix factorisation (NMF). Once the item (or user) similarity is computed, either the R user-item matrix, or the user (tag cloud) profile obtained from U or I are used to predict the recommendations for a user. For instance, [24] presents a framework based on the three matrices, U, I and R, to recommend web pages (based on http://del.icio.us data). Also, [25] uses matrix I to improve the accuracy results of the recommendations, after combining I with the results obtained by classic collaborative filtering. Levy applies Latent Semantic Analysis (that is; SVD and cosine similarity in the reduced space) to compute and visualise artist similarity derived from tags gathered from *last.fm* [26].

Finally, it is worth mentioning that inverting either U or I matrices, one can also compute tag similarity. Tag similarity have many usages in recommendation and search engines. For instance, tag synonym detection can be used for query expansion, or tag suggestion when annotating the content.

Using the 3-order tensor (instead of decomposing the tensor in bidimensional matrices) is the second approach to mine the data, and provide recommendations. The available techniques are (high-order) extensions of SVD and NMF. HOSVD is a higher order generalisation of matrix SVD for tensors, and Non-negative Tensor Factorisation (NTF) is a generalisation of NMF.

In [27], the authors apply HOSVD to a music dataset (user-artists-tags) taken from *last.fm*. Their results show significant improvements in terms of the effectiveness measured through precision and recall. Yanfei et al. present a similar method using bookmarking data from *del.icio.us* [28]. They apply SVD on the R matrix, compute cosine distance among the users (to find the neighbours), and then apply classic CF user-based recommendation (see Sec. 2.5.2). The authors could improve the results over a CF approach based on SVD and cosine similarity (e.g. Latent Semantic Analysis).

2.5.4.4 Limitations of Social Tagging

One of the main limitations of social tagging is the coverage. On the one hand, it is quite common that only the most popular items are described by several users, creating a compact description of the item. On the other hand, long tail items usually do not have enough tags to characterise them. This makes the recommendation process very difficult, specially to promote these unknown items.

Another issue is that without being constrained to a controlled vocabulary, tags present the following problems: polysemy (I *love* this song, versus this song is about *love*), synonymy (*hip-hop*, *hiphop*, and *rap*), and usefulness of the personal tags to derive similarity among users or items (e.g. *seen live*, or *to check*). These issues make more difficult to mine and extract useful relationships among the items and the users.

Tag sparsity is another issue. In some domains, some tags are widely used (e.g. *rock* or *pop*, in the music domain), whereas other tags are rarely applied (e.g. *gretsch guitar*). A biased distribution of the terms has also consequences when exploiting social tagging data.

Last but not least, any recommender system that relies on user explicit input can be attacked, or vandalised. Users can deliberately mistag some items in order to provoke an undesired effect in the recommendations (see Paris Hilton example presented in Sec. 3.3.3).

2.5.5 Hybrid Methods

The main purpose of a hybrid method is to achieve a better recommendations by combining some of the previous stand-alone approaches. Most commonly, collaborative filtering is combined with other techniques. There are several methods to integrate different approaches into a hybrid recommender. Some of the methods that Burke defines are [29]:

- *Weighted*. A hybrid method that combines the output of separate approaches using, for instance, a linear combination of the scores of each recommendation technique.
- *Switching*. The system uses some criterion to switch between recommendation techniques. One possible solution is that the system uses a technique, and if the results are not confident enough, it switches to another technique to improve the recommendation process.
- *Mixed*. In this approach, the recommender does not combine but expand the description of the data sets by taking into account the users' ratings and the description of the items. The new prediction function has to cope with both types of descriptions.
- *Cascade*. The cascade involves a step by step process. In this case, a recommendation technique is applied first, producing a coarse ranking of items. Then, a second technique refines or re-rank the results obtained in the previous step.

A hybrid method can alleviate some of the drawbacks that suffer a single technique.

2.6 Factors Affecting the Recommendation Problem

2.6.1 Novelty and Serendipity

The novelty factor is a very important aspect of the recommendation problem. It has been largely acknowledged that providing obvious recommendations can decrease user satisfaction [2, 30]. Obvious recommendations have two practical disadvantages: users who are interested in those items could probably already know them, and secondly, managers in stores (i.e. experts of the items' domain) do not need any recommender to tell them which products are popular overall.

Although, obvious recommendations do have some value for new users. Users like to receive some recommendations they already are familiar with [31]. This is related with the *Find credible recommender* use case (see Sec. 2.2). Yet, there is a trade-off between the desire for novel versus familiar recommendations. A high novelty rate might mean, for a user, that the quality of the recommendation is poor, because the user is not be able to identify most of the items in the list of recommendations. However, by providing explanations (transparency) of the recommendations, the user can feel that is a credible recommender. Thus, the user can be more open to receive novel, justified, recommendations.

Another important feature, closely related with novelty is the serendipity effect. That is the good luck in making unexpected and fortunate discoveries. A recommender should help the user to find a surprisingly interesting item that she might not be able to discover otherwise. Recommendations that are serendipitous are also novel and relevant for a user.

2.6.2 Explainability

Explainability (or transparency) of the recommendations is another important element. Giving explanations about the recommended items could increase user trustiness and loyalty of the system, and also her satisfaction.

A recommender should be able to explain to the user why the system recommends the list of *top-K* items [32]. Herlocker et al. present an experimental evidence that shows that providing explanations can improve the acceptance of those recommender systems based on collaborative filtering [33]. Actually, giving explanations about why the items were recommended is as important as the actual list of recommended items. Tintarev and Masthoff summarise the possible aims for providing recommendations. These are: transparency, scrutability, trust, effectiveness, persuasiveness, efficiency, and satisfaction. They also stress the importance of personalising the explanations to the user [34].

2.6.3 Cold Start Problem

As already mentioned, the cold start problem of a recommender (also known as the *learning rate curve*, or the *bottleneck problem*) happens when a new user (or a new item) enters into the system [35]. On the one hand, cold start is a problem for users that just signed-up, because the system does not have enough information about them. If the user profile initialisation is empty (see Sec. 2.4.1), she has to dedicate some time using the system before getting some useful recommendations. On the other hand, when a new item is added to the collection, the system should have enough information to be able to recommend this item to users.

2.6.4 Data Sparsity and High Dimensionality

Data sparsity is an inherent property of the dataset. With a relative large number of users and items, the main problem is the low coverage of the users' interaction with the items. A related factor is the high dimensionality of the dataset, that consists of many users and items.

There are some methods, based on dimensionality reduction, that alleviate data sparsity and high dimensionality of the dataset. Singular Value Decomposition (SVD), and Non-negative Matrix Factorisation (NMF) [16, 36, 37] are the two most used methods in recommendation. Takács et al. present in [38] several matrix factorisation algorithms, and evaluate the results against the Netflix Prize dataset.

2.6.5 Coverage

The coverage of a recommender measures the percentage of the items in the collection over which the system, or make recommendations. A low coverage of the domain might be less valuable to users, as it limits the space of possible items to recommend. Moreover, this feature is important for the *Find all good items* use case (see Sec. 2.2). Also, a low coverage of the collection can be very frustrating for the users, and clearly affects the novelty and serendipity factors.

2.6.6 Trust

Trust-aware recommender systems determine which users are reliable, and which are not. Trust computational models are needed, for instance, in user-based CF to rely on the user's neighbours.

In [39], the authors present two computational models of trust and show how they can be readily incorporated into CF. Furthermore, combining trust and classic CF can improve the predictive accuracy of the recommendations. Massa et al. emphasise the "web of trust" provided by every user. They use the "web of trust" to propagate trust among users, and also use it to alleviate the data sparsity problem. An empirical evaluation shows that using trust information improves the predictive accuracy, as well as the coverage of the recommendations [40, 41].

2.6.7 Attacks

Recommender systems can be attacked in various ways, degrading the quality of the recommendations. For instance, sybil attacks try to subvert the system by creating a large number of sybil identities in order to gain a lot of influence in the system. These attacks can promote or demote some particular items.

A related problem is the user profile injection, where a malicious user fakes some of her data creating an incongruent profile. Then, using a particular rating pattern that highly rates a set of target items, and then rating other items so they become similar. This is known as *shilling* attacks [42].

Another problem is deliberate mistagging. That is when a group of users tag an item using a (malicious) tag. This behavior can affect the performance of social-based recommenders.

2.6.8 Temporal Effects

Temporal effects are found in both items and users. On the one hand, the timestamp of an item (e.g. when the item was added to the collection) is an important factor for the recommendation algorithm. The prediction function can take into account the *age* of an item. A common approach is to treat the older items as less relevant than the new ones, promoting new items that are continuously added in the collection.

On the other hand, the system has to decide which items from a user profile are taken into account when computing the recommendations. Should the system use all the information of a profile, or only the latest activity? Depending on the criteria, it might change the output recommendations.

In this context, [43] presents the recommendation problem as a sequential optimisation problem. It is based on Markov decision processes (MDP). MDP uses the long-term effects of the recommendations, but it is also configurable to use only the last *k*-actions of a user. The main problem, though is the computationally complexity of the algorithm, which makes it unusable for large datasets.

Temporal effects are also found in the manner how users rate items. Users' ratings can also vary over time. When a user has to rate a given set of items over a period of time, the ratings provided by her are not always consistent [5].

2.6.9 Understanding the Users

Modelling user preferences, including psychographic information, is another challenging problem. Psychographic variables include attributes related with personality; such as attitudes, interests, or lifestyles. It is not straightforward to encode all this information and use it in the recommender system. This problem is similar in Information Retrieval (IR) systems, where users have to express their needs via a keyword-based query. There is a loss of information when a user is formulating a query using a language that the machine can understand and process. When dealing with user profiles and sensitive personal information, privacy is an important aspect.

2.7 Summary

This chapter has presented and formalised the recommendation problem. The main components of a recommender are users and items. Based on the user preferences and the exploitation of a user profile, a recommender can solve the problem of recommending a list of items to a user, or a list of like-minded users. There are several factors that affect the recommendation problem, and in this book we emphasise the novelty one. We believe that this is an important topic that deserves to be analysed in depth.

To recap, Table 2.1 presents the main elements involved in the recommendation problem, that is user profiling (generation, maintenance, and adaptation), and the recommendation methods (matching items—or users—to a user, and the filtering methods to match them).

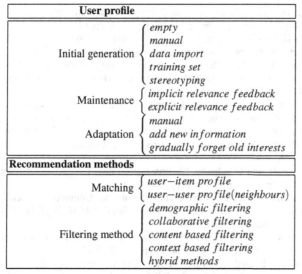

User profile		
Initial generation	$\left\{\begin{array}{l}\end{array}\right.$	*empty* *manual* *data import* *training set* *stereotyping*
Maintenance	$\left\{\begin{array}{l}\end{array}\right.$	*implicit relevance feedback* *explicit relevance feedback*
Adaptation	$\left\{\begin{array}{l}\end{array}\right.$	*manual* *add new information* *gradually forget old interests*
Recommendation methods		
Matching	$\left\{\begin{array}{l}\end{array}\right.$	*user−item profile* *user−user profile(neighbours)*
Filtering method	$\left\{\begin{array}{l}\end{array}\right.$	*demographic filtering* *collaborative filtering* *content based filtering* *context based filtering* *hybrid methods*

Table 2.1 Summary of the elements involved in the recommendation problem.

References

1. B. Sarwar, G. Karypis, J. Konstan, and J. Reidl, "Item-based collaborative filtering recommendation algorithms," in *WWW'01: Proceedings of 10th International Conference on World Wide Web*, (New York, NY), pp. 285–295, 2001.
2. J. L. Herlocker, J. A. Konstan, L. G. Terveen, and J. T. Riedl, "Evaluating collaborative filtering recommender systems," *ACM Transactions on Information Systems*, vol. 22, no. 1, pp. 5–53, 2004.
3. M. Montaner, B. Lopez, and J. L. de la Rosa, "A taxonomy of recommender agents on the internet," *Artificial Intelligence Review*, vol. 19, pp. 285–330, 2003.
4. W. Hill, L. Stead, M. Rosenstein, and G. Furnas, "Recommending and evaluating choices in a virtual community of use," in *Proceedings of SIGCHI Conference on Human Factors in Computing Systems*, (New York, NY), pp. 194–201, 1995.
5. X. Amatriain, J. M. Pujol, N. Tintarev, and N. Oliver, "Rate it again: Increasing recommendation accuracy by user re-rating," in *Proceedings of the ACM Conference on Recommender Systems*, (New York, NY), ACM, 2009.
6. G. Webb and M. Kuzmycz, "Feature based modelling: A methodology for producing coherent, consistent, dynamically changing models of agents' competencies," in *User Modeling and User-Adapted Interaction*, vol. 5, no. 2, pp. 117–150, 1996.
7. K. McCarthy, M. Salamó, L. Coyle, L. McGinty, B. Smyth, and P. Nixon, "Group recommender systems: A critiquing based approach," in *Proceedings of the 11th International Conference on Intelligent User Interfaces*, (New York, NY), pp. 267–269, ACM, 2006.
8. Y.-L. Chen, L.-C. Cheng, and C.-N. Chuang, "A group recommendation system with consideration of interactions among group members," *Expert Systems with Applications*, vol. 34, no. 3, pp. 2082–2090, 2008.
9. E. Rich, "User modeling via stereotypes," *Cognitive Science: A Multidisciplinary Journal*, vol. 3, no. 4, pp. 329–354, 1979.
10. M. J. Pazzani, "A framework for collaborative, content-based and demographic filtering," *Artificial Intelligence Review*, vol. 13, no. 5–6, pp. 393–408, 1999.

11. D. Goldberg, D. Nichols, B. M. Oki, and D. Terry, "Collaborative filtering to weave and information tapestry," *Communications of the ACM*, vol. 35, pp. 61–70, December 1992.

12. U. Shardanand, "Social information filtering for music recommendation," Master's thesis, Massachussets Institute of Technology, Cambridge, MA, September 1994.

13. U. Shardanand and P. Maes, "Social information filtering: Algorithms for automating "word of mouth"," in *Proceedings of SIGCHI Conference on Human Factors in Computing Systems*, (Denver, CO), ACM, 1995.

14. P. Resnick, N. Iacovou, M. Suchak, P. Bergstorm, and J. Riedl, "Grouplens: An open architecture for collaborative filtering of netnews," in *Proceedings of ACM 1994 Conference on Computer Supported Cooperative Work*, (Gaithersburg, MD), pp. 175–186, ACM, 1994.

15. P. Resnick and H. R. Varian, "Recommender systems," *Communications of the ACM*, vol. 40, no. 3, pp. 56–58, 1997.

16. Y. Koren, R. Bell, and C.Volinsky, "Matrix factorization techniques for recommender systems," *IEEE Computer*, vol. 42, no. 8, pp. 30–37, 2009.

17. M. Claypool, A. Gokhale, T. Miranda, and P. Murnikov, "Combining content-based and collaborative filters in an online newspaper," *Proceedings of ACM SIGIR Workshop on Recommender Systems*, (Berkeley, CA), 1999.

18. C. Avery and R. Zeckhauser, "Recommender systems for evaluating computer messages," *Communications of the ACM*, vol. 40, no. 3, pp. 88–89, 1997.

19. M. J. Salganik, P. S. Dodds, and D. J. Watts, "Experimental study of inequality and unpredictability in an artificial cultural market," *Science*, vol. 311, pp. 854–856, February 2006.

20. G. D. Abowd, A. K. Dey, P. J. Brown, N. Davies, M. Smith, and P. Steggles, "Towards a better understanding of context and context-awareness," in *Proceedings of the 1st International Symposium on Handheld and Ubiquitous Computing*, (London, UK), pp. 304–307, Springer, 1999.

21. R. Kosala and H. Blockeel, "Web mining research: A survey," *SIGKDD Explorations*, vol. 2, pp. 1–15, 2000.

22. B. Mobasher, R. Cooley, and J. Srivastava, "Automatic personalization based on web usage mining," *Communications of the ACM*, vol. 43, no. 8, pp. 142–151, 2000.

23. G. Salton and M. J. McGill, *Introduction to Modern Information Retrieval*. New York, NY: McGraw-Hill, Inc., 1986.

24. A.-T. Ji, C. Yeon, H.-N. Kim, and G. Jo, "Collaborative tagging in recommender systems," in *Australian Conference on Artificial Intelligence*, vol. 4830 of *Lecture Notes in Computer Science*, pp. 377–386, Springer, 2007.

25. K. H. L. Tso-Sutter, L. B. Marinho, and L. Schmidt-Thieme, "Tag-aware recommender systems by fusion of collaborative filtering algorithms," in *Proceedings of the ACM Symposium on Applied Computing*, (New York, NY), pp. 1995–1999, ACM, 2008.

26. M. Levy and M. Sandler, "A semantic space for music derived from social tags," in *Proceedings of the 8th International Conference on Music Information Retrieval*, (Vienna, Austria), 2007.

27. P. Symeonidis, M. Ruxanda, A. Nanopoulos, and Y. Manolopoulos, "Ternary semantic analysis of social tags for personalized music recommendation," in *Proceedings of 9th International Conference on Music Information Retrieval*, (Philadelphia, PA), 2008.

28. Y. Xu, L. Zhang, and W. Liu, "Cubic analysis of social bookmarking for personalized recommendation," *Frontiers of WWW Research and Development*, pp. 733–738, 2006.

29. R. Burke, "Hybrid recommender systems: Survey and experiments," *User Modeling and User–Adapted Interaction*, vol. 12, no. 4, pp. 331–370, 2002.

30. S. M. McNee, J. Riedl, and J. A. Konstan, "Being accurate is not enough: How accuracy metrics have hurt recommender systems," in *Computer Human Interaction. Human Factors in Computing Systems*, (New York, NY), pp. 1097–1101, ACM, 2006.

31. K. Swearingen and R. Sinha, "Beyond algorithms: An HCI perspective on recommender systems," in *ACM SIGIR. Workshop on Recommender Systems*, vol. 13, nos. 5–6, pp. 393–408, 2001.

32. R. Sinha and K. Swearingen, "The role of transparency in recommender systems," in *CHI - Extended Abstracts on Human Factors in Computing Systems*, (New York, NY), pp. 830–831, ACM, 2002.

33. J. L. Herlocker, J. A. Konstan, and J. Riedl, "Explaining collaborative filtering recommendations," in *Proceedings of the ACM Conference on Computer Supported Cooperative Work*, (New York, NY), pp. 241–250, ACM, 2000.

34. N. Tintarev and J. Masthoff, "Effective explanations of recommendations: User-centered design," in *Proceedings of the ACM Conference on Recommender Systems*, (Minneapolis, MN), pp. 153–156, ACM, 2007.

35. D. Maltz and K. Ehrlich, "Pointing the way: Active collaborative filtering," in *Proceedings of SIGCHI Conference on Human Factors in Computing Systems*, (New York, NY), pp. 202–209, ACM/Addison-Wesley Publishing Co., 1995.

36. P. Paatero and U. Tapper, "Positive matrix factorization: A non-negative factor model with optimal utilization of error estimates of data values," *Environmetrics*, vol. 5, no. 2, pp. 111–126, 1994.

37. D. D. Lee and H. S. Seung, "Learning the parts of objects by non-negative matrix factorization," *Nature*, vol. 401, pp. 788–791, October 1999.

38. G. Takács, I. Pilászy, B. Németh, and D. Tikk, "Investigation of various matrix factorization methods for large recommender systems," in *Proceedings of the 2nd KDD Workshop on Large Scale Recommender Systems and the Netflix Prize Competition*, (Las Vegas, NV), 2008.

39. J. O'Donovan and B. Smyth, "Trust in recommender systems," in *Proceedings of the 10th International Conference on Intelligent User Interfaces*, (New York, NY), pp. 167–174, ACM, 2005.

40. P. Massa and B. Bhattacharjee, "Using trust in recommender systems: An experimental analysis," in *Proceedings of iTrust International Conference*, pp. 221–235, 2004.

41. P. Massa and P. Avesani, "Trust-aware recommender systems," in *Proceedings of the ACM Conference on Recommender Systems*, (New York, NY), pp. 17–24, ACM, 2007.

42. P.-A. Chirita, W. Nejdl, and C. Zamfir, "Preventing shilling attacks in online recommender systems," in *Proceedings of the 7th Annual ACM International Workshop on Web Information and Data Management*, (New York, NY), pp. 67–74, ACM, 2005.

43. G. Shani, R. I. Brafman, and D. Heckerman, "An MDP-based recommender system," *Journal of Machine Learning Research*, vol. 6, pp. 453–460, 2002.

Chapter 3
Music Recommendation

This chapter focuses on the recommendation problem in the music domain. Section 3.1 presents some common use cases in music recommendation. After that, Sec. 3.2, discusses user profiling and modelling, and how to link the elements of a user profile with the music concepts. Then, Sec. 3.3 presents the main components to describe the musical items, that are artists and songs. The existing music recommendation methods (collaborative filtering, content, context-based, and hybrid) and the pros and cons of each approach are presented in Sect. 3.4. Finally, Sec. 3.5 summarises the work presented, and provides some links with the remaining chapters of the book.

3.1 Use Cases

The main task of a music recommendation system is to propose interesting music, consisting of a mix of known and unknown artists—as well as the available tracks—given a user profile. Most of the work done in music recommendation focuses on presenting to a user a list of artists, or creating an ordered sequence of songs (a personalised playlist). Yet, there are other interesting scenarios. For instance, providing recommendations for a group of users, in a particular context [1, 2]. That is, an automatic DJ that selects music to please as much people in the party as possible. Or, proposing background music for a restaurant, given some constraints such as the type of music (ambient, relaxed, or only instrumental, etc.), the lyrics' language (e.g. only Italian songs in a pizzeria), as well as other particularities that the restaurant might impose. Another scenario could be a very specialised advanced search system for a music producer. The producer might need to search for a specific bassline or bass sound, that should fit into the whole song. A music recommender should be able to assist both the boss of the restaurant and the producer.

Ò. Celma, *Music Recommendation and Discovery*,
DOI 10.1007/978-3-642-13287-2_3, © Springer-Verlag Berlin Heidelberg 2010

3.1.1 Artist Recommendation

According to the general model presented in Chap. 2 (see Fig. 2.1), artist recommendation follows the user-item matching, were items are recommended to a user
according to her profile. However, artist recommendation should involve a broader
experience with the user, more than presenting a list of relevant artists, plus some
accompanying metadata. In this sense, there is a lot of music related information on
Internet: music performed by "unknown" —long tail—artists that can suit perfectly
for new recommendations, new music releases, related news, concerts listings, album reviews, mp3-blogs, podcasts, t-shirts, and a long etcetera.

Indeed, music websites syndicate (part of) their web content—noticing the user
about new releases, artists' related news, upcoming gigs, etc.—in the form of RSS
(Really Simple Syndication) feeds. For instance, the *iTunes Music Store*[1] provides
an RSS feed generator[2] updated once a week, that publishes all the new releases
of the week. A music recommendation system could take advantage of all these
publishing services.

3.1.2 Playlist Generation

Cunningham et al. make the distinction between a playlist and a *mix*. In a mix, the
order of the songs is important, whilst a playlist focuses more on a desired emotional
state, or acts as a background to an activity (while working, while reading, while
jogging, etc.) [3].

Playlist generation is an important application in music recommendation, as it
allows users to listen to the music as well as provide immediate feedback, so the system can react accordingly. There are several ways to automatically create a playlist;
shuffle (i.e random), based on a given seed song—or artist—or based on a user-
profile (including her like-minded neighbours). With regard to the available music,
there are two main modes of playlist generation: (i) using tracks drawn from the
users own collection (ii) using tracks drawn from the *celestial jukebox*.[3]

3.1.2.1 Shuffle, Random Playlists

Interestingly enough, some experiments have been carried out to investigate serendipity in random playlists. Nowadays, shuffle is still the usual way to generate playlists
on personal computers and portable music players. A study of serendipity through

---

[1] http://www.apple.com/itunes

[2] http://ax.itunes.apple.com/rss, accessed April, 17th 2009

[3] For example music that comes from the *Playdar* music content resolver service http://www.
playdar.org/

shuffle playlists is presented in [4]. The authors argue that shuffle can invest new meanings to a particular song. It provides opportunities for unexpected re-discoveries, and also in some cases re-connects songs with old memories. Although, serendipity can be achieved by creating more personalised and elaborated playlists, rather than purely based on random choices.

3.1.2.2 Personalised Playlists

Radio-a-la-carte (personalised playlists) is another way to propose music to a user. In this case, music is selected in terms of the user preferences, within a particular context. The user can also provide feedback (e.g. *Skip this song*, *More like this*, etc.) according to her taste and the actual listening context.

3.1.3 Neighbour Recommendation

The goal of neighbour recommendation is to find like-minded people. Neighbour similarity is based on the user–user profile matching presented in Fig. 2.1. Once a user is set in a neighbourhood, she can discover music through her neighbours, or simply be part of that community (or cluster) and interact with them.

One of the main advantages of creating neighbourhoods is that a user can ex-plore and discover music via her neighbours. Also, it promotes the creation of tight communities, connecting people that share similar interests.

3.2 User Profile Representation

Music is an important vehicle for telling other people something relevant about our personality, history, etc. Musical taste and music preferences are affected by several factors, including demographic and personality traits. It seems reasonable to think that music preferences and personal aspects—such as: age, gender, origin, occupa-tion, musical education, etc.—can improve music recommendation [5].

User modelling has been studied for many years. Yet, extending a user profile with music related information has not been largely investigated. Indeed, it is an interesting way to communicate with other people, and to express their music pref-erences.[4]

---

[4] Nowadays, it is very common to embed to a webpage a small widget that displays the most recent tracks a user has played.

3.2.1 Type of Listeners

The *Phoenix 2* UK Project from 2006 summarises the four degrees of interest in
music, or type of listeners [6]. This study is based on the analysis of different type
of listeners, with an age group ranging from 16 to 45. The classification, depicted in
Fig. 3.1, includes:

- *Savants*. Everything in life seems to be tied up with music. Their musical knowledge is very extensive. As expected, they only represent 7% of the 16–45 age
  group.
- *Enthusiasts*. Representing 21% of the 16–45 age group, for the enthusiasts music
  is a key part of life but is also balanced by other interests.
- *Casuals*. Music plays a welcome role, but other things are far more important.
  They represent 32% of the 16–45 age group.
- *Indifferents* would not lose much sleep if music ceased to exist. Representing
  40% of the 16–45 age group, they are a predominant type of listeners of the
  whole population.

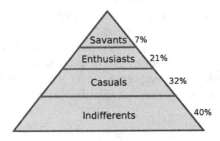

Fig. 3.1 The four type of music listeners: savants, enthusiasts, casuals, and indifferents. Each type
of listener needs different type of recommendations.

Each type of listener needs different type of recommendations. Savants do not
really need popular recommendations, but risky and clever ones. They are the most
difficult listeners to provide recommendations, because they are very exigent. Enthusiasts appreciate a balance between interesting, unknown, recommendations and
familiar ones. Casuals and indifferents (72% of the population) do not need any
complicated recommendations. Probably, popular, mainstream music that they can
easily identify would fit their musical needs. Thus, a recommender system should
be able to detect the type of user and act accordingly.

3.2.2 Related Work

3.2.2.1 Context in Music Perception

Lesaffre et al. reveal in [7] that music perception is affected by contextual information, and this context depends on each user. The study explores the dependencies of demographic and musical background for different users in an annotation experiment. Subject dependencies are found for age, music expertise, musicianship, taste and familiarity with the music. The authors propose a semantic music retrieval system based on fuzzy logic. The system incorporates the annotations of the experiment, and music queries are done using semantic descriptors. The results are returned to the user, based on her profile and preferences. Again, one of the main conclusions of their research is that music search and retrieval systems should distinguish between the different categories of users.

3.2.2.2 Subjective Perception of Music Similarity

In [8], the authors present a music recommendation engine based on user's perceived similarity. User similarity is defined as a combination of timbre, genre, tempo, year and mood. The system allows users to define the weights for personalised playlist generation.

Sotiropoulos et al. state that different users assess music similarity via different feature sets, which are in fact subsets of some set of objective features. They define a subset of features, for a specific user, using relevance feedback and a neural network for incremental learning [9].

Going one step beyond, the work presented in [10] allows users to defining their own semantic concepts (e.g. happy, blue, morning-music, etc.), providing some instances—sound excerpts—that characterise each concept. The system adapts, then, can adapt to these user's concepts and it predicts (using audio content-based similarity) the labels for the newly added songs. This process is also known as *autotagging*. The system can also generate a playlist based on one or more user's concepts.

3.2.2.3 The User in the Community

A single user profile can be extended taking into account her interaction with the community of peers. Tracking social network activity allows a system to infer user preferences. Social networks have a big potential not only for the social interactions among the users, but also to exploit recommendations based on the behaviour of the community, or even to provide group-based recommendations.

In [11], the authors present a recommendation framework based on social filtering. The user profile consists on static and dynamic social aspects. The dynamic aspect includes the interaction with other users, and the relationships among them (e.g. duration, mutual watchings of web pages, common communications, etc.).

Analysing this information, the authors present novel ways of providing social filtering recommendations.

Another example is the *Bluetuna* system. *Bluetuna* is a "socialiser engine" based on sharing user preferences for music [12]. *Bluetuna* allows users to share musical tastes with other people who are (physically) near by. The application runs on bluetooth enabled mobile phones. The idea is to select those users that have similar musical tastes, facilitating the meeting process.

Firan et al. create tag-based user profiles using social tagging information derived from the collective annotation [13]. Once a user profile is described using a tag cloud, the authors present several approaches to compute music recommendations. The results show an accuracy improvement using tag-based profiles over traditional collaborative filtering at song level.

3.2.2.4 Privacy Issues

When dealing with user profiles and sensitive personal information, privacy is an important aspect. In [14], the authors present some results about the acquisition, storage and application of sensitive personal information. There is a trade-off between the benefits of receiving personalised music recommendations and the lost of privacy. The factors that influence disclosing sensitive personal information are:

- the purpose of the information disclosure,
- the people that get access to the information,
- the degree of confidentiality of the sensitive information, and
- the benefits they expect to gain from disclosing it.

3.2.3 User Profile Representation Proposals

As noted in the previous section, music recommendation is highly dependent on the type of user. Also, music is an important vehicle for conveying to others something relevant about our personality. User modelling, then, is a crucial step in understanding user preferences.

However, in the music recommendation field, there have been few attempts to explicitly extend user profiles by adding music related information. The most relevant (music-related) user profile representation proposals are: the User modelling for Information Retrieval Language, the MPEG-7 standard that describes user preferences, and the Friend of a Friend (FOAF) initiative (hosted by the Semantic Web community). The complexity, in terms of semantics, increases with each proposal. The following sections present these three approaches.

3.2.3.1 User Modelling for Information Retrieval (UMIRL)

The UMIRL language, proposed by [15], allows one to describe perceptual and qualitative features of the music. It is specially designed for music information retrieval systems. The profile can contain both demographic information and direct information about the music objects: favourite bands, styles, songs, etc. Moreover, a user can add her definition of a perceptual feature and his meaning, using music descriptions. For instance: "a *romantic piece* has a slow tempo, lyrics are related with *love*, has a soft intensity, and the context to use this feature is while having a special dinner with user's girlfriend".

The representation they proposed uses the XML syntax, without any associated schema nor document type definition to validate the profiles. Listing 3.1 shows a possible user profile:

```
<user>
 <generalbackground>
   <name>Joan Blanc</name>
   <education>MsC</education>
   <citizen>Catalan</citizen>
 </generalbackground>
 <musicbackground>
   <education>none</education>
   <instrument>guitar</instrument>
 </musicbackground>
 <musicpreferences>
   <genre>rock</genre>
   <album>
     <title>To bring you my love</title>
     <artist>P.J. Harvey</artist>
   </album>
 </musicpreferences>
</user>
```

Listing 3.1 Example of a user profile in UMIRL.

This proposal is one of the first attempts in the Music Information Retrieval community. The main goal was to propose a representation format, as a way to interchange profiles among systems, though, it lacks formal semantics to describe the meaning of their descriptors and attributes. To cope with this limitation, the following section presents an approach using the MPEG-7 standard.

3.2.3.2 MPEG-7 User Preferences

MPEG-7, formally named Multimedia Content Description Interface, is an ISO/IEC standard developed by the Moving Picture Experts Group (MPEG). The main goal of the MPEG-7 standard is to provide structural and semantic description mechanisms for multimedia content. The standard provides a set of description schemes (DS) to describe multimedia assets. In this section, we only focus on the descriptors that describes user preferences of multimedia content. An in-depth description of the whole standard appears in [16].

User preferences in MPEG-7 includes content filtering, searching and browsing preferences and usage history, which represents the user history of interaction with multimedia items, can be denoted too. Filtering and searching preferences include the user preferences regarding classification (i.e. country of origin, language, available reviews and ratings, reviewers, etc.) and creation preferences. The creation preferences describes the creators of the content (e.g. favourite singer, guitar player, composer, and music bands). Also, it allows one to define a set of keywords, location and a period of time. Using a preference value attribute, a user can express positive (likes) and negative (dislikes) preferences for each descriptor. The following example shows a partial user profile definition, stating that this user likes the album *To bring you my love* from *P.J. Harvey*:

```
<UserPreferences>
 <UserIdentifier protected="true">
  <Name xml:lang="ca">Joan Blanc</Name>
 </UserIdentifier>
 <FilteringAndSearchPreferences>
  <CreationPreferences>
   <Title preferencValue="8">To bring you my love</Title>
   <Creator>
    <Role>
     <Name>Singer</Name>
    </Role>
    <Agent xsi:type="PersonType">
     <Name>
      <GivenName>Polly Jean</GivenName>
      <FamilyName>Harvey</FamilyName>
     </Name>
    </Agent>
   </Creator>
   <Keyword>dramatic</Keyword>
   <Keyword>fiery</Keyword>
   <DatePeriod>
    <TimePoint>1995-01-01</TimePoint>
    <Duration>P1825D</Duration>
   </DatePeriod>
  </CreationPreferences>
 </FilteringAndSearchPreferences>
</UserPreferences>
```

Listing 3.2 Example of a user profile in MPEG-7.

MPEG-7 usage history is defined following the usage history description scheme. *UsageHistory DS* contains a history of user actions. It contains a list of actions (play, play-stream, record, etc.), with an associated observation period. The action has a program identifier (an identifier of the multimedia content for which the action took place) and, optionally, a list of related links or resources.

Tsinaraki et al. present a way to overcome some of the limitations of describing user preferences in MPEG-7 [17]. They argue that there is still a lack of semantics when defining user preferences, as the whole MPEG-7 standard is based on XML Schemas. For example, filtering and search preferences allow one to specify a list of textual keywords, without being related to a taxonomy or ontology. Their implementation is integrated into a framework, based on an upper ontology that covers the MPEG-7 multimedia description schemes. This upper ontology uses the Web Ontology Language (OWL)[5] notation. So it does the next proposal, based on the Friend of a Friend initiative.

3.2.3.3 FOAF: User Profiling in the Semantic Web

The Friend of a Friend project provides conventions and a language "to tell" a machine the type of things a user says about herself in her homepage. Friend of a Friend is based on the RDF/XML vocabulary. As we noted before, the knowledge held by a community of "peers" about music is also a source of valuable metadata. Friend of a Friend nicely allows one to easily relate and connect people.

Friend of a Friend profiles include demographic information (name, gender, age, sex, nickname, homepage, depiction, web accounts, etc.) geographic (city and country, geographic latitude and longitude), social information (relationship with other persons), pyschographic (i.e. user's interests) and behavioural (usage patterns). There are some approaches to model music preferences music taste in a Friend of a Friend profile.

The simplest way to show interest for an artist is shown in the following example:

```
<foaf:interest>
  rdf:resource="http://www.pjharvey.net"
  dc:title="P.J._Harvey" />
```

Listing 3.3 Example of a user interest using FOAF.

The Semantic Web approach facilitates the integration of different ontologies. Listing 8.5 shows how to express that a user likes an artist, using the general *Music Ontology* proposed in [18].

```
<foaf:interest>
 <mo:MusicArtist rdf:about="http://musicbrainz.org/artist/ca37
     -...fc">
  <mo:discogs rdf:resource="http://www.discogs.com/artist/PJ+
     Harvey"/>
  <foaf:img rdf:resource="http://ec2.images-amazon.com/images/P/
     B00852Q....jpg"/>
  <foaf:homepage rdf:resource="http://pjharvey.net/"/>
  <foaf:name>P.J. Harvey</foaf:name>
  <mo:wikipedia rdf:resource="http://en.wikipedia.org/wiki/
     PJ_Harvey"/>
 </mo:MusicArtist>
</foaf:interest>
```

Listing 3.4 Example of a user interest using FOAF, and the music ontology to describe the artist.

To conclude this section, Example 3.5 shows a complete Friend of a Friend profile. This profile contains demographic and geographic information, as well as user's interests —with a different level of granularity when describing the artists.

```
<rdf:RDF
  (XML namespaces here)
```

---

[5] http://www.w3.org/TR/owl-features/

```
>
<foaf:PersonalProfileDocument rdf:about="">
 <foaf:maker rdf:resource="#me"/>
 <foaf:primaryTopic rdf:resource="#me"/>
 <admin:generatorAgent
    rdf:resource="http://foafing-the-music.iua.upf.edu"
 />
 <admin:errorReportsTo
    rdf:resource="mailto:ocar.celma@upf.edu"/>
</foaf:PersonalProfileDocument>
<foaf:Person rdf:ID="me">
 <foaf:nick>ocelma</foaf:nick>
 <foaf:dateOfBirth>04-17</foaf:dateOfBirth>
 <foaf:gender>male</foaf:gender>
 <foaf:based_near geo:lat='41.401' geo:long='2.159' />
 <foaf:holdsAccount>
  <foaf:OnlineAccount>
   <foaf:accountName>ocelma</foaf:accountName>
   <foaf:accountServiceHomepage
      rdf:resource="http://last.fm"/>
  </foaf:OnlineAccount>
 </foaf:holdsAccount>
 <foaf:mbox_sha1sum>ce24ca...a1f0</foaf:mbox_sha1sum>
 <foaf:interest>
  <foaf:Document rdf:about="http://www.gretsch.com">
   <dc:title>Gretsch guitars</dc:title>
  </foaf:Document>
 </foaf:interest>
 <foaf:interest>
  <mo:MusicArtist rdf:about="http://musicbrainz.org/artist/ca37
      -...fc">
   <mo:discogs rdf:resource="http://www.discogs.com/artist/PJ+
      Harvey"/>
   <foaf:img rdf:resource="http://ec2.images-amazon.com/images/P
      /B00852Q....jpg"/>
   <foaf:homepage rdf:resource="http://pjharvey.net/"/>
   <foaf:name>P.J. Harvey</foaf:name>
   <mo:wikipedia rdf:resource="http://en.wikipedia.org/wiki/
      PJ_Harvey"/>
  </mo:MusicArtist>
 </foaf:interest>
 </foaf:Person>
</rdf:RDF>
```

Listing 3.5 Example of a user's FOAF profile

This approach, based on the Friend of a Friend notation, is the one used in one of the two prototypes, named *Foafing the music*, presented in Chap. 8 (Sect. 8.2).

3.3 Item Profile Representation

Now we describe the representation and modelling of music items. That is, the main elements that describe artists and songs. First we introduce, in Sec. 3.3.1, the Music Information Plane (MIP). MIP defines the different levels of complexity and abstraction to describe music assets. After that, we classify these semantic descriptions using the music knowledge classification (editorial, cultural and acoustic metadata) proposed by Pachet in [19].

3.3.1 The Music Information Plane

In the last twenty years, the signal processing and computer music communities have developed a wealth of techniques and technologies to describe audio and music content at the lowest (close to the signal) level of representation. However, the gap between these low-level descriptors and the concepts that music listeners use to relate it with a music collection (the so-called "semantic gap") is still, to a large extent, waiting to be bridged.

Due to the inherent complexity when describing multimedia objects, a layered approach with different levels of granularity is needed. In the multimedia field and, specially, in the music field we foresee three levels of abstraction: low-level basic features, mid-level semantic features, and high-level human understanding. The first level includes physical features of the objects, such as the sampling rate of an audio file, as well as some basic features like the spectral centroid of an audio frame, or even the predominant chord in a sequential list of frames. A mid-level of abstraction aims at describing concepts such as a guitar solo, or tonality information (e.g. key and mode) of a track. Finally, the higher level should use reasoning methods or semantic rules to retrieve, for instance, several audio files with "similar" guitar solos over the same key.

We describe the music information plane in two dimensions. One dimension considers the different media types that serve as input data (audio, text and image). The other dimension is the level of abstraction in the information extraction process of this data. Figure 3.2 depicts the music information plane.

The input media types, in the horizontal axis, include data coming from: audio (music recordings), text (lyrics, editorial text, press releases, etc.) and image (video clips, CD covers, printed scores, etc.). On the other side, for each media type there are different levels of information extraction (in the vertical axis). The lowest level is located at the signal features. This level lays far away from what an end-user might find meaningful. Anyway, it is the basis to describe the content and to produce more elaborated descriptions of the media objects on top of that. This level includes basic audio features (such as: energy, frequency, mel frequency cepstral coefficients, or even the predominant chord in a sequential list of frames), or basic natural language processing for the text media. At the mid-level (the content objects level), the information extraction process and the elements described are a bit closer to the end-user. This level includes description of musical concepts (such as a guitar solo, or tonality information—e.g. key and mode—of a music title), or named entity recognition for text information. Finally, the higher-level, includes information related with human beings and their interaction with music knowledge. This level could use inference methods and semantic rules to retrieve, for instance, several audio files with similar guitar solos over the same key. Also, in this level, there is the user and her social

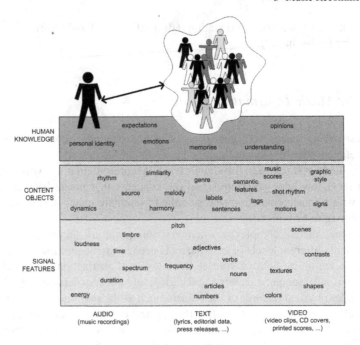

Fig. 3.2 The music information plane. The horizontal axis includes the input media types. The vertical axis represents the different levels of information extraction for each media type. At the *top*, a user interacts with the music content and the social network of users.

relationships with a community of users. Figure 3.2 depicts the music information plane.

Nonetheless, the existing semantic gap between concept objects and human knowledge makes it more difficult for a music recommender system. This semantic gap has many consequences to music understanding and music recommendation. Yet, there are some open questions, such as: which are the music elements that makes a person feel certain emotions, or to evoke some particular memories? How is a personal identity linked with music? Only a multi-modal approach, that takes into account as much elements from MIP as possible, would be able to (partly) answer some of these questions. Neither pure bottom-up nor top-down approaches can lead to bridge this gap. We foresee, then, an approximation in both ways: users need to interact with the content to add proper (informal) semantics (e.g. via tagging), and also content object descriptions must be somehow *understandable* by the users.

Pachet classifies the music knowledge management in three categories [19]. The three categories are: editorial, cultural and acoustic metadata. This classification allows one to create meaningful descriptions of music, and to exploit these descriptions to build music recommendation systems. In the following sections, we depict each category in the music information plane.

3.3.2 Editorial Metadata

Editorial metadata (EM) consists of information manually entered by an editor. Usually, the information is decided by an expert or a group of experts. Figure 3.3 depicts the relationship between editorial metadata and the music information plane.

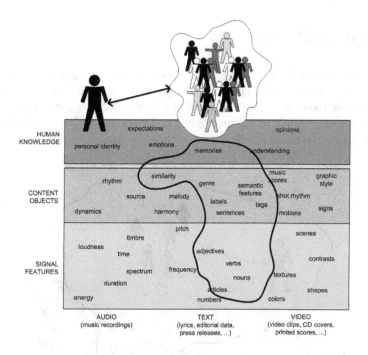

Fig. 3.3 Editorial metadata and the music information plane.

Editorial metadata includes simple creation and production information (e.g. the song *C'mon Billy*, written by P.J. Harvey in 1995, was produced by John Parish and Flood, and the song appears as track number 4, on the album "To bring you my love"). Editorial metadata includes, in addition, artist biography, genre information, relationships among artists, etc. Thus, editorial information is not necessarily objective. It is usual the case that different experts cannot agree in assigning a concrete genre to a song or to an artist. Even more difficult is a common consensus of a taxonomy of musical genres.

The scope of Editorial metadata is rather broad. Yet, it usually refers to these items: the creator (or author) of the content, the content itself, and the structure of the content. Regarding the latter, editorial metadata can be fairly complex. For example, an opera performance description has to include the structure of the opera. It is divided in several acts. Each act has some scenes. In a given scene, there is a soprano singing an *Aria* piece, and many musicians playing. It has lyrics to sing, and these can be in different languages (sung in Italian, but displayed in English), etc.

In terms of music recommendation, EM conforms the core for non content-based methods for music recommenders.

3.3.3 Cultural Metadata

Cultural metadata (CM) is defined as the information that is implicitly present in huge amounts of data. This data is usually gathered from Internet; via weblogs, forums, music radio programs, etc. CM has a clear subjective component as it is based on the aggregation of personal opinions. Figure 3.4 depicts the relationship between cultural metadata and the music information plane.

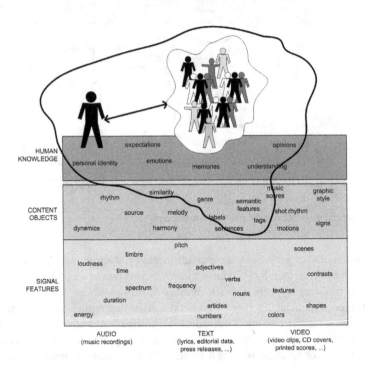

Fig. 3.4 Cultural metadata and the music information plane.

3.3.3.1 Web-MIR Techniques to Describe Artists

Web Music Information Retrieval (Web-MIR) is a recent field of research in the MIR community. Web-MIR focuses on the analysis and exploitation of cultural information. So far, Web-MIR performances close to *classic* content-based approaches are reported on artist genre classification and artist similarity [20–22]. Yet, it is not clear how Web-MIR methods can deal with long tail content.

The origins of Web-MIR can be found in the earlier work of Whitman et al. [20, 23]. They describe artists using a list of weighted terms. To gather artist related terms, they query a general search engine with the name of the artist. To limit the size of the page results, they add some keywords to the query, such as "music" and "review". From the retrieved pages, the authors extract unigrams, bigrams and noun phrases. In [23], Whitman uses an unsupervised method for music understanding, using the power spectral density estimate over each 5 seconds of audio. Then, it keeps the semantically dimensions that contain the most significant meanings. Similarly, in [24] Baumann et al. improved this approach by filtering irrelevant content of the web pages (e.g. adverts, menus, etc.). The description of an artist is conformed by the terms with the highest normalised TF-IDF value. That includes the most relevant nouns, adjectives and simple phrases, as well as un-tagged unigrams and bigrams.

In [25], the authors present different ways to describe artists using web data, based on co-occurrence analysis between an artist and the labels. The set of labels are previously defined, and conform a corpus of music related terms (e.g. genres, instruments, moods, etc.). The three methods they use are: Pagecount-based mapping (PCM), Pattern-based mapping (PM), and Document-based mapping (DM). PCM uses the total number of hits retrieved by *Google* search engine. However, some terms appear more often than others (e.g. *pop*, or *rock* versus *cumbia*). So, they provide a normalised version, inspired by Pointwise mutual information (see Sect. 2.5.4). Pattern-based mapping uses a set of predefined English phrase patterns. For example "*(genre)* artists such as *(artist)*". An instance of the pattern could be: "*Country* artists such as". This way, the method can retrieve the most prominent Country artists. Table 3.1 shows the results for the Country style pattern.[6] Finally, document-based mapping analyses the content of the top-k pages returned by *Google*. That is, the algorithm downloads the most representative pages, according to the query, and then counts the music related terms found in the k pages. It is worth noting that these three methods can also be used not only to characterise the artists, but to compute artist similarity.

Similar work based on co-occurrences is presented in [21, 22]. In [21], the authors define artist similarity as the conditional probability of an artist that occurs on a web page that was returned as response to querying another artist. In [22], the authors focus on artist genre classification, using three different genre taxonomies. An artist assignment to a genre is considered as a special form of co-occurrence analysis. An evaluation performed on a small dataset shows an accuracy of over 85%. Related to this, Zadel and Fujinaga investigate artist similarity using *Amazon* and

---

[6] The query was performed on September, 9th 2008, using *Google* search engine. The results were manually analysed, and only the first page (top-10 results) was used.

Artist	# occurrences
Garth Brooks	2
Hank Williams	2
Shania Twain	2
Johnny Cash	1
Crystal Gayle	1
Alan Jackson	1
Webb Pierce	1
Carl Smith	1
Jimmie Rodgers	1
Gary Chapman	1

Table 3.1 A list of prominent Country artists obtained using Pattern-based matching on *Google* (on September, 9th 2008). The results were manually analysed, and only the first page (top-10 results) was used.

Listmania! APIs, and then *Google* to refine the results, using artist co-occurrences in webpages [26].

One of the main drawbacks of Web-MIR is the polysemy of some artists' names, such as *Kiss*, *Bush*, *Porn* [27]. This problem is partially solved by the same authors, in [28]. Based on TF-IDF, they penalise the terms with high DF, that is the terms that appear in lots of documents.

A common drawback of all the previous approaches is the high dimensionality of the datasets. To avoid this problem, Pohle et al. use Non-negative Matrix Factorisation to reduce the dimensionality of the artist-term matrix [29]. They also use a predefined vocabulary of music terms, and analyse the content of the top-100 web pages related to each artist. To get the most relevant pages, they use a similar approach as Whitman and Lawrence [20]. The original matrix contains all the terms applied to the artists, using TF-IDF weights. This matrix is decomposed into 16 factors, or "archetypical" concepts using non-negative matrix factorisation. Then, each artist is described by a 16-dimensional vector. After that, a music browser application allows users to navigate the collection by adjusting the weights of the derived concepts, and also can recommend similar artists using cosine distance over the artists' vectors.

Another source to derive artist or song similarity is based on the analysis of available (or manually created) playlists on the web. Automatic playlists based on song co-occurrences typically use web data mining techniques to infer song similarity. That is crawling public playlists, and computing song or artist co-occurrence from this data. For instance, Baccigalupo et al. [30] analysed artists co-occurrences using a set of more than 1 million playlists from the *MyStrands* web. Pachet also computes artist and song co-occurrences from radio sources using a big database of CD compilations, extracted from *CDDB* [31]. Cunningham et al. state that playlists contain a lot of context, and only humans are able to interpret it (e.g. "music about my holidays back in 1984") [3]. According to a user survey done by the same authors, only 25% of the mixes are organised using content related information, such as artist, genre or style. The rest is based on contextual information.

3.3.3.2 Collecting Ground Truth Data

An important aspect when trying to evaluate similarity metrics using cultural meta-data is the creation of reliable ground truth data. Different proposals are presented in [19, 24, 32, 33]. The problem of gathering ground-truth for music similarity evaluation is outlined in [34]. In [33], Geleijnse et al. create a *dynamic* ground truth for artist tagging and artist similarity. The idea is to adapt to the dynamically changing data being harvested by social tagging (e.g. from *last.fm*), instead of defining a static and immutable ground truth.

To summarise, Turnbull et al. present five different ways to collect annotations at artist (or song) level. These approaches are [35]:

- mining web documents,
- harvesting social tags,
- autotagging audio content,
- deploying annotation games, and
- conducting a survey

Cultural information, based on Web-MIR and social tagging techniques, is the basis for context-based music recommenders. Section 3.4.2 presents the main ideas to exploit cultural information, and use it to provide music recommendations.

3.3.3.3 Social Tagging Vandalism: The *Paris Hilton — Brutal Death Metal* Case

When using data from the wisdom of the crowds one needs to pay attention to users' intentional misuse or mistag of the items. That is, social tagging spam, or vandalism. In May 2007, while Paul Lamere and myself were preparing the Music Recommendation Tutorial for the ISMIR conference, he found out a problem with Paris Hilton artist tags in *last.fm*.[7] A group of users were deliberately tagging her with *Brutal Death Metal* tag.[8] Of course, this affected the recommendations of the system, where all of a sudden one could hear a *death metal* song in a Paris Hilton playlist (and the other way around!). Figure 3.5 shows a *last.fm* screenshot[9] with top artists tagged with *Brutal Death Metal*. Top-1 artist is Paris Hilton. Indeed, looking at her raw tag counts in Table 3.2, we see that top-1 tag is *Brutal Death Metal*. Also, there are some other descriptive—so to say—tags such as *atainwptiosb*,[10] *Your ears will bleed*, or *the worst thing ever to happen to music*. It is clear that some users were having fun with her.

Social tagging spam is a problem for any music recommender system that relies on this type of data to derive artist (or track) similarity. We outline some possible solutions to post-process the artist's tag list, and "clean" it:

---

[7] http://www.last.fm/music/Paris+Hilton

[8] http://blogs.sun.com/plamere/entry/the_1_brutal_death_metal

[9] Screenshot taken on May, 23rd 2007

[10] Acronym of *all things annoying in the world put together into one stupid b*tch*

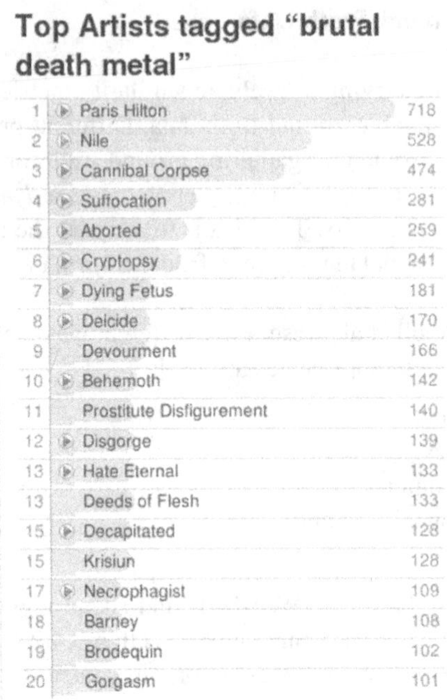

Top Artists tagged "brutal death metal"

1	▶ Paris Hilton	718
2	▶ Nile	528
3	▶ Cannibal Corpse	474
4	▶ Suffocation	281
5	▶ Aborted	259
6	▶ Cryptopsy	241
7	▶ Dying Fetus	181
8	▶ Deicide	170
9	Devourment	166
10	▶ Behemoth	142
11	Prostitute Disfigurement	140
12	▶ Disgorge	139
13	▶ Hate Eternal	133
13	Deeds of Flesh	133
15	▶ Decapitated	128
15	Krisiun	128
17	▶ Necrophagist	109
18	Barney	108
19	Brodequin	102
20	Gorgasm	101

Fig. 3.5 *Last.fm* screenshot with top artists tagged with *Brutal Death Metal*. Screenshot taken on May, 23rd 2007.

Tagger reliability. The system might try to reduce the influence of untrusted taggers (tagger reliability). Some questions that might help identifying those users are:

- Does the tagger listen to the music they are tagging (e.g. Paris Hilton music)?
- Does the tagger often use the tags that they are applying? (e.g. *atainwptiosb* or *the worst thing ever to happen to music*)
- Does anyone else use those (potential spam) tags to other artists?

Once the system identifies them, it can act accordingly. For instance, diminishing the tagging effect of these users.

Tag Clustering. The idea here is to compare other Paris Hilton tags (such as *pop*, *female vocalist, sexy* or *guilty pleasure*) against *Brutal Death metal* tag. This way, we can see whether all these tags are correlated, and belong to the same semantic cluster. Table 3.3 shows different examples of tag similarity, using a *last.fm* dataset with 84,838 artists and 187,551 distinct tags. We apply Latent Semantic Analysis (LSA). That is, to compute Singular Value Decomposition (SVD) over the artist-tag-frequency matrix, reducing it to 100 factors or dimensions (see Sec. 2.5.2). After that, we use cosine similarity to compute tag similarity. There is no semantic similarity between *Brutal Death Metal* and *pop, female vocalist, sexy* or *guilty*

Tag	Raw count
Brutal Death Metal	1,145
atainwptiosb	508
Crap	290
Pop	287
Officially Sh*t	248
Sh*t	143
Your ears will bleed	140
emo	120
whore	103
in prison	98
female vocalist	80
whore untalented	79
Best Singer in the World	72
sexy	50
the worst thing ever to happen to music	47
b*tch	42
dance	41
Guilty Pleasures	40
Death Metal	30
Female	29
Slut	29

Table 3.2 *Last.fm* raw tag counts for Paris Hilton artist. Accessed on May, 23rd 2007, via its API v1.0.

pleasure. On the other hand, *pop* and *guilty pleasure*, or *sexy* and *female vocalist* are much more similar (see the last two examples in Table 3.3). Figure 3.6 shows a denodogram, based on the LSA cosine similarity from Table 3.3. The resulting hierarchical clustering shows how *Brutal Death Metal* belongs to a different, isolated, cluster than the rest of the tags.

LSA_{cosine} **similarity(Tag$_1$, Tag$_2$)**
sim(brutal death metal, pop) = 0.055
sim(brutal death metal, female vocalist) = 0.034
sim(brutal death metal, sexy) = 0.066
sim(brutal death metal, guilty pleasure) = 0.027
sim(pop, guilty pleasure) = 0.399
sim(sexy, female vocalist) = 0.759

Table 3.3 LSA cosine similarity between *Brutal Death Metal* and some other Paris Hilton tags.

Furthermore, one can get the tags from other artists correctly tagged with *Brutal Death metal* (e.g. *death metal*, *extreme metal*, *gore metal* or *grindcore*), and see whether these tags also appear in (or, in general, co-occur with) other Paris Hilton tags. The underline idea here is to detect artist tags that could be outliers (i.e. potential spam tags).

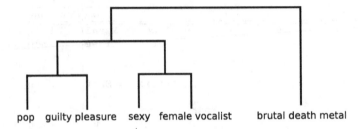

pop   guilty pleasure   sexy   female vocalist   brutal death metal

Fig. 3.6 Dendogram for some Paris Hilton tags, including *Brutal Death Metal*, using the cosine similarity results from Table 3.3.

Listening habits. Taking into account how many users listen to both Paris Hilton and any other *Brutal Death Metal* band can give us an insight on how related are those artists. That is, to compute the (co-occurrence analysis or collaborative filtering) similarity between her and some other prominent death metal artists.

Tag	*Last.fm* relevance
Pop	100
Female Vocalists	28
Dance	18
American	14
Sexy	13
Brutal Death Metal	11
rnb	8
female vocalist	8
female	7
00s	6
Guilty Pleasure	6
California	5
emo	4
Crap	3
Reggae	3
awful	3
party	3
underrated	2
Best Singer in the world	2
atainwptiosb	2

Table 3.4 *Last.fm* normalised tags for Paris Hilton after doing a post-processing to clean tag spam.

To conclude this section, Table 3.4 presents the solution proposed by *last.fm*. It shows a list of tags for Paris Hilton, after a post-processing and cleaning algorithm. Their solution was partly inspired on Koutrika's work [36]. We can see how now *Brutal Death Metal* is not anymore at position top-1. It still appears in the list, but with a lower (normalised) relevance of 11 out of 100.

3.3.4 Acoustic Metadata

The last category of semantic music description is acoustic metadata. Acoustic metadata extracts features from the audio, using content-based analysis. Semantic acoustic descriptors are the basis for content-based music recommenders (see Sect. 3.4.3). Figure 3.7 depicts the relationship between acoustic metadata and the music information plane.

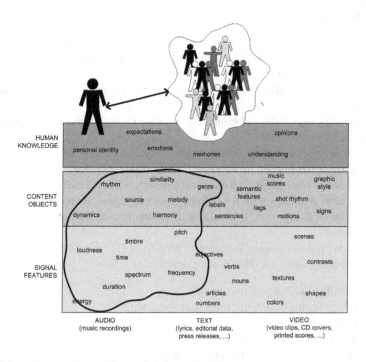

Fig. 3.7 Acoustic metadata and the music information plane.

Most of the current music content processing systems operating on complex audio signals are mainly based on computing low-level signal features. These features are good at characterising the acoustic properties of the signal, returning a description that can be associated to a texture. A more general approach consists in describing music content according to several "musical facets" (i.e. rhythm, harmony, melody, timbre, etc.) by incorporating higher-level semantic descriptors. Semantic descriptors can be computed directly from the audio signal combining signal processing, machine learning, and musical knowledge. The following sections are devoted to outlining some relevant music description facets.

3.3.4.1 Low-Level Timbre Descriptors

To describe the audio it is very usual to decompose the audio signal with spectral and temporal features. Spectral features are considered more robust to polyphonic and complex textures. The signal is segmented into (overlapping) frames, generally from 10 to 100 ms with, say, 50% overlap. For each frame, a feature vector is computed. Now, we briefly present some of the most important low-level timbre features to describe the audio signal. Most of them are based on the Short-Time Fourier Transform (STFT):

- *Spectral Centroid* is a concept adapted from psychoacoustics and music cognition. Spectral Centroid is the mean value of the STFT amplitude spectrum. It measures the average frequency, weighted by amplitude, of the spectrum.
- *Spectral Flatness* is the ratio between the geometrical mean and the arithmetical mean of the spectrum magnitude.
- *Spectral Skewness* is the 3rd order central moment, and it gives indication about the shape of the spectrum.
- *Spectral Kurtosis* is the 4th order central moment. It measures whether the data are peaked or flat relative to a normal (Gaussian) distribution.
- *Zero-Crossing Rate* (ZCR) is a temporal descriptor defined as the number of time domain zero-crossings within a defined region of signal, divided by the number of samples of that region. It measures, then, the rate of sign-changes along the audio signal.
- *Mel Frequency Cepstrum Coefficients (MFCCs)* [37] are widely used in speech recognition applications. MFCC are calculated as follows:

  1. Divide the audio signal into frames.
  2. For each frame, obtain the amplitude spectrum.
  3. Take the logarithm.
  4. Convert to Mel spectrum.
  5. Take the discrete cosine transform (DCT).

  Step 4 calculates the log amplitude spectrum on the Mel scale. The Mel transformation is based on human perception experiments. Then, step 5 takes the DCT of the Mel spectra.

The *bag-of-frames* timbre approach consists in modelling the audio signal using a statistical distribution of the audio features, on short-time audio segments. Audio features are then aggregated together using simple statistics (e.g. mean and variance), or modelled as a Gaussian Mixture Model (GMM). However, as pointed out by Aucouturier and Pachet in [38], a timbre representation based on MFCCs and GMMs tend to create hubs. These are songs that are irrelevantly close to every other songs.

Furthermore, similarity methods solely based on describing timbre information tend to find similar pieces that belong to different music genres. It is very unlikely that a user will love both a *Franz Schubert*'s piano sonata, and a *Meat Loaf* ballad just because the two contain a prominent piano melody. In the following sections we present other music facets that complement timbre audio features.

3.3.4.2 Instrumentation

Extracting truly instrumental information from music, as pertaining to separate instruments or types of instrumentation implies classifying, characterising and describing information which is buried behind many layers of highly correlated data. Given that the current technologies do not allow a sufficiently reliable separation, most research work has concentrated on the characterisation of the "overall" timbre or "texture" of a piece of music as a function of low-level signal features. This approach implied describing mostly the acoustical features of a given recording, gaining little knowledge about its instrumental contents [39].

Even though it is not yet possible to fully separate the different contributions and "lines" of the instruments, there are some simplifications that can provide useful descriptors (e.g. lead instrument recognition, solo detection). The recognition of idiosyncratic instruments, such as percussive ones, is another valuable simplification. Given that the presence, amount and type of percussion instruments are very distinctive features of some music genres percusive information can be exploited to provide other natural partitions to large music collections. Herrera et al. define semantic descriptors such as the percussion index, or the percussion profile [40]. Although they can be computed after doing (simple) source separation, reasonable approximations can be achieved using simpler sound classification approaches that do not attempt separation [41]. Additionally, [42] presents an instrument identification, of mono-instrumental music, using line spectral frequencies (LSF) and k-means classifier.

3.3.4.3 Rhythm

In its most generic sense, rhythm refers to all of the temporal aspects of a musical work, whether represented in a score, measured from a performance, or existing only in the perception of the listener [43]. In the literature the concept of "automatic rhythm description" groups many applications as diverse as tempo induction, beat tracking, rhythm quantisation, meter induction and characterisation of timing deviations, to name a few. Many of these different aspects have been investigated, from the low-level onset detection, to the characterisation of music according to rhythmic patterns.

At the core of automatic rhythmic analysis lies the issue of identifying the start, or onset time, of events in the musical data. As an alternative to standard energy-based approaches, another methodologies have recently appeared: a method that works solely with phase information [44], or that are based on predicting the phase and energy of signal components in the complex domain [45], greatly improving results for both percussive (and tonal) onsets. However, there is more to rhythm than the absolute timings of successive musical events. For instance, [46] proposes a general model of beat tracking, based on the use of comb-filtering techniques on a continuous representation of "onset emphasis". This method was expanded to combine this general model with a context-dependent model by including a state space switching model. This improvement has been shown to significantly improve upon previous results, in particular with respect to maintaining a consistent metrical level and preventing phase switching between off-beats and on-beats.

Furthermore, the work done by Gouyon and Dixon ([47, 48]) demonstrates the use of high-level rhythmic descriptors for genre classification of recorded audio. An example is a tempo-based classification showing the high relevance of this feature while trying to characterise dance music [47]. However, this approach is limited by the assumption that, given a musical genre, the tempo of any instance is among a very limited set of possible tempi. To address this, [48] use bar-length rhythmic patterns for the classification of dance music. The method dynamically estimates the characteristic rhythmic pattern on a given musical piece, by a combination of beat tracking, meter annotation and a k-means classifier. Genre classification results are greatly improved by using these high-level descriptors, showing the relevance of musically-meaningful representations for Music Information Retrieval tasks. Dannenberg presents in [49] a holistic approach toward automated beat tracking, taking into account music structure. Last but not least, for a complete overview of the state of the art on rhythmic description the reader is referred to [43].

3.3.4.4 Harmony

The harmony of a piece of music can be defined by the combination of simultaneous notes, or chords; the arrangement of these chords along time, in progressions; and their distribution, which is closely related to the key or tonality of the piece. Chords, their progressions, and the key are relevant aspects of music perception that can be used to accurately describe and classify music content.

Harmonic based retrieval has not been extensively explored before. A successful approach at identifying harmonic similarities between audio and symbolic data was presented in [50]. It relied on automatic transcription, a process that is partially effective within a highly constrained subset of musical recordings (e.g. mono-timbral, no drums or vocals, small polyphonies). To avoid such constraints [51] adopts the approach where describes the harmony of the piece, without attempting to estimate the pitch of notes in the mixture. Avoiding the transcription step allows to operate on a wide variety of music. This approach requires the use of a feature set that is able to emphasise the harmonic content of the piece, such that this representation can be exploited for further, higher-level, analysis. The feature set of choice is known as a Chroma or Pitch Class Profile, and they represent the relative intensity of each of the twelve semitones of the equal-tempered scale.

Gómez et al. present in [52] an approach of the tonality estimation by correlating chroma distributions with key profiles, derived from music cognition studies. Results show high recognition rates for a database of classical music. The studies done in [53] have also concentrated on chord estimation based on chroma features, using tuning, and a simple template-based model of chords. Recognition rates of over 66% were found for a database of recorded classical music, though the algorithm is being used also with other musical genres. A recent development includes the generation of a harmonic representation using a Hidden Markov Model, initialised and trained using musical theoretical and cognitive considerations [54]. This methodology has already shown great promise for both chord recognition and structural segmentation. For a deeper overview of harmonic and tonality description see [55].

3.3.4.5 Structure

Music structure refers to the ways music materials are presented, repeated, varied or confronted along a piece of music. Strategies for doing that are artist, genre and style-specific (i.e. the *A–B* themes exposition, development and recapitulation of a sonata form, or the *intro–verse–chorus–verse–chorus–outro* of "pop music"). Detecting the different structural sections, the most repetitive segments, or even the least repeated segments, provide powerful ways of interacting with audio content based on summaries, fast-listening and musical gist-conveying devices, and on-the-fly identification of songs.

The section segmenter developed by Ong and Herrera in [56] extracts segments that roughly correspond to the usual sections of a pop song or, in general, to sections that are different (in terms of timbre and tonal structure) from the adjacent ones. The algorithm first performs a rough segmentation with the help of change detectors, morphological filters adapted from image analysis, and similarity measurements using low-level descriptors. It then refines the segment boundaries using a different set of low-level descriptors. Complementing this type of segmentation, the most repetitive musical pattern in a music file can also be determined by looking at self-similarity matrices in combination with a rich set of descriptors including timbre and tonality (i.e. harmony) information.

3.3.4.6 Intensity

Subjective intensity, or the sensation of energeticness we get from music, is a concept commonly and easily used to describe music content. Although intensity has a clear subjective facet, Sandvold et al. hypothesised that it could be grounded on automatically extracted audio descriptors. Inspired by the findings of Zils and Pachet in [57], Sandvold et al. created a model of subjective intensity built from energy and timbre low-level descriptors extracted from the audio data [58]. They have proposed a model that decides among 5 labels (ethereal, soft, moderate, energetic, and wild), with an estimated effectiveness of nearly 80%. The model has been developed and tested using several thousands of subjective judgements.

3.3.4.7 Genre

Music genres are connected to emotional, cultural and social aspects, and all of them influence our music understanding. The combination of these factors produce a personal organization of music which is, somehow, the basis for (human) musical genre classification. Indeed, musical genres have different meanings for different people, communities, and countries [59, 60].

The use of musical genres has been deeply discussed by the MIR community. A good starting point is the review by McKay [61]. The authors suggest that musical genres are an inconsistent way to organize music. Yet, musical genres remain a very effective way to describe and tag artists. Broadly speaking, there are two complementary approaches when defining a set of genre labels: (i) the definition of a controlled vocabulary by a group of experts or musicologists, and (ii) the collaborative effort of a community (social tagging). The goal of the former approach is the creation of a list of terms, organised in a hierarchy. A hierarchy includes the relationships among the terms; such as hyponymy. The latter method, social tagging, is a less formal bottom-up approach, where the set of terms emerge during the (manual) annotation process.

Music genre classification is a classic MIR problem. The setup consists of predicting one (or more) genre labels from the audio. Most of the approaches use machine learning methods to train a classifier per genre, using a combination of audio features. Early work in automatic genre classsification is presented by Tzanetakis and Cook [62]. The authors use timbre related features (Spectral Centroid, Spectral Rolloff, Spectral Flux, and MFCC) as well as rhythm features based on the beat histogram. A complete state of the art on music genre classification is presented in [63].

3.3.4.8 Mood

According to Juslin and Laukka, people listen to music mostly to change their emotional state [64]. When dealing with such subjective question, we are faced with several issues. The first one is the emotion representation paradigm. There exist two main mood representations from psychology: a dimensional, continuous model, and a categorical, discrete list. As advised by Juslin et al., one should consider few categories when building a ground truth that maximises the agreement between people [65].

Extracting moods from the audio is a very challenging task. In the Music Information Retrieval field only recent work, using exclusively audio content, deals with this problem. Most of these approaches use machine learning techniques, training a classifier with some selected audio features [66–68]. Laurier et al. observe from the audio analysis (using timbre, rhythm and tonal descriptors) the correlation between psychological emotions and musical features [68]. For example a fast tempo (onset rate feature) and major tonality song is classified as *happy*, while a slow tempo and minor tonality might correspond to a *sad* emotion. Given the limitations to classify music emotions using only audio content-based features, recent approaches include hybrid methods combining contextual (e.g mood tags) and audio content information [69].

3.3.4.9 Tools and Resources

We briefly present in Table 3.5 some audio processing tools and libraries. This software is widely used in the Music Information Retrieval community. It is not our goal to detail the pros and cons of each tool. We list them here as starting point for those readers interested in extracting audio features, and computing audio content-based similarity on top of them.

Name	Language	Link
MA Toolbox	Matlab	http://www.ofai.at/~elias.pampalk/ma/
MIR Toolbox	Matlab	http://users.jyu.fi/~lartillo/
Marsyas	C++	http://marsyas.info/
CLAM	C++	http://clam-project.org/
SMIRK	ChucK	http://smirk.cs.princeton.edu/
Echo nest	Web API	http://developer.echonest.com/pages/overview

Table 3.5 A list of tools to extract audio features from the signal.

Other libraries that extract low-level descriptors such as MFCC/GMM are: Music Browser (in Matlab),[11] and Auditory Toolbox,[12] also in Matlab. Furthermore, "The Tools We Use" webpage[13] compiles a list of resources that the MIR community widely uses.

3.4 Recommendation Methods

In this section, we present the music recommendation methods to match user preferences (see Sec. 3.2) with the artist and music description (presented in the previous Sec. 3.3).

---

[11] http://www.jj-aucouturier.info/projects/mir/

[12] http://cobweb.ecn.purdue.edu/~malcolm/interval/1998-010/

[13] http://www.music-ir.org/evaluation/tools.html

3.4.1 Collaborative Filtering

Collaborative filtering (CF) techniques have been largely applied in the music domain. CF makes use of the editorial and cultural information. Early research was
based on explicit feedback, based on the ratings about songs or artists. Yet, tracking
user listening habits has become the most common way in music recommendation.
In this sense, CF has to deal with implicit feedback (instead of explicit ratings).

3.4.1.1 Explicit Feedback

Ringo, described in [70], is the first music recommender based on collaborative filtering and explicit feedback (ratings). The author applies user-based CF approach
(see Sec. 2.5.2). Similarity among users is computed with Pearson normalised correlation (see Eq. 2.3). Then, the recommendations are computed as the mean of the
ratings done by the similar users of the active user (see Eq. 2.6).

Racofi (Rule Applying COllaborative FIltering) approach combines collaborative
filtering based on ratings, and a set of logic rules based on Horn clauses [71]. The
rules are applied after the ratings have been gathered. The five rating dimensions
they define are: impression, lyrics, music, originality, and production. The objective
of the rules is to prune the output of the collaborative filtering, and promote the
items that the user will be most familiar with. Anderson et al. exemplifies a rule
[71]:

> If a user rates 9 the originality of an album by artist X then the predicted originality rating,
> for this user, of all other albums by artist X is increased by a value of 0.5.

These kind of rules implicitly modify the ratings that a user has done previously.
The *Indiscover* music recommender system[14] implements this approach, as well as
the Slope One collaborative filtering method, presented in [72].

3.4.1.2 Implicit Feedback

Implicit feedback in the music domain is usually gathered from users' listening
habits. The main drawback is that the value that a user *assigns* to an item is not
always in a predefined range (e.g. from 1..5 or *like it/hate it*). Instead, the interaction
between users and items is usually described by songs she listens to, or the total
playcounts. Thus, the system can only track (implicit) positive feedback. Negative
feedback cannot be gathered. Only when users explicitly rate the content, the range
of values include both positive and negative feedback (e.g. from 1..5 stars, where 1
means a user does not like the item, 3 indifference, and 5 she loves it).

Furthermore, recommendations are usually performed at artist level (unless the
system generates a playlist for that user), whilst listening habits are at song level.
In this case, an aggregation process—from song plays to artist total playcounts—is
needed. To use CF with implicit feedback at artist level, there are different options:

---

[14] http://www.indiscover.net

- Convert the implicit data into a binary user-artist matrix M. Non-zero cells mean that the user has listened to the artist at least once.
- Transform the implicit data into a normalised matrix. Instead of assigning 0/1 to a cell, the value can denote how much a user listens to the artist. E.g. [0..5], where 5 denotes that she listens to a lot the artist, 1 means only from time to time, and 0 never. This matrix has a more fine-grained description of the user listening habits than the previous, binary, normalisation.
- Normalise each row (users), so that the sum of the row entries equal 1. This option, then, describes the artist probability distribution of a user.
- Do not normalise the matrix. The matrix contains for each user and artist ($M_{i,j}$) the total playcounts.

In any case, after the dataset is represented in the user-artist matrix, one can apply any CF methods with explicit feedback (presented in Sec. 2.5.2).

It is common that the user's listening habits distribution is skewed to the right, so it shows a heavy-tailed curve. That is, a few artists have lots of plays in the user profile, and the rest of artists have much less playcounts. Figure 3.8 depicts the listening habits of a user in terms of total playcounts. The horizontal axis contains her top-50 artists, ranked by the total plays (i.e. artist at position 1 has 238 playcounts).

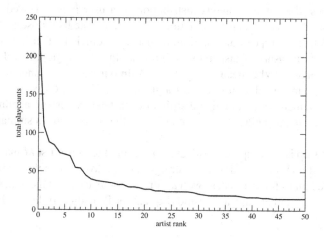

Fig. 3.8 A user listening habits represented with frequency distribution of playcounts per artist in the user's profile.

Then, we compute the complementary cumulative distribution of artist plays in the user profile. Artists located in the top 80–100% of the distribution get a score of 5, artists in the 60–80% range get a 4, and so on (until the artists with less playcounts, in the 0–20% range, which get assigned a 1). The rest of the M_i cells have value 0. Figure 3.9 shows the complementary cumulative distribution of the artist playcounts from Fig. 3.8.

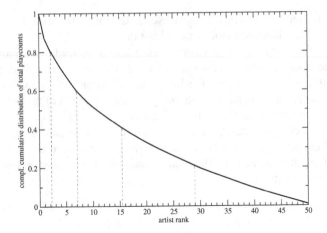

Fig. 3.9 User listening habits from Fig. 3.8 represented with the complementary cumulative distribution. Top-1 and 2 artists receive a score of 5. Artists at position 3..7 have a score of 4, artists in 8..15 a 3, and so on.

Sometimes, the listening habits distribution of a user is not skewed, but very homogeneous (a small standard deviation value, and a median close to the mean value). To detect this type of distribution, we use the coefficient of variation, CV. CV is a normalised measure of dispersion of a probability distribution, that divides the standard deviation by the mean value, $CV = \frac{\sigma}{\mu}$. In our case, the standard deviation of plays by the mean value of plays, for a given user. Then, if $CV \leq 0.5$ we do not use the complementary cumulative distribution. Instead, we assign a value of 3 to all the user artists, meaning that all the artists in the profile have a similar number of plays.

The next step is to compute artist similarity using the user-artist M matrix. Once the normalisation process is done, it is straightforward to compute the average value of normalised plays for an artist, as well as for a user—in case that the item similarity measure to use is either adjusted cosine (Eq. 2.2) or Pearson correlation (Eq. 2.3).

3.4.1.3 An Example

We have done some experiments with data obtained from *last.fm*. The dataset contains the top-artists playcounts for more than 500,000 users, with a total of around 30 million ⟨*user, artist, playcount*⟩ triples. To clean the list of artists, we only use those artists that have a *Musicbrainz*[15] ID, and also that at least 10 users listened to them once or more. After the cleaning process, we get a list of around 95,000 distinct artists. To apply CF, we transformed the listening habits dataset to a user-artist matrix M. $M_{i,j}$ represents the number of times user i has listened to artist j. To nor-

---

[15] http://www.musicbrainz.org

malise the matrix we followed the second approach; computing the complementary cumulative distribution. That is to assign a range value $[0..5]$ in $M_{i,j}$ from the $\langle user_i, artist_j, playcount \rangle$ (as shown in Fig. 3.9).

We present two concrete examples of item-similarity using Pearson correlation, (Eq. (2.3)) and conditional probability (Eq. 2.4) from user-artist matrix M. Table 3.6 (left) shows the top-10 similar artists of *The Dogs d'Amour*,[16] whilst the right column shows the results obtained using conditional probability similarity.

The Dogs d'Amour	Similarity$_{Pearson}$	The Dogs d'Amour	Similarity$_{Cond.Prob.}$
Los Fabulosos Cadillacs	0.806	Guns n' Roses	0.484
Electric Boys	0.788	Aerosmith	0.416
Lillian Axe	0.784	AC/DC	0.379
Michael Jackson	0.750	Led Zeppelin	0.360
Ginger	0.723	Metallica	0.354
The Decemberists	0.699	Alice Cooper	0.342
The Byrds	0.667	Mötley Crüe	0.341
Zero 7	0.661	David Bowie	0.335
Rancid	0.642	Red Hot Chili Peppers	0.334
The Sonics	0.629	The Beatles	0.334

Table 3.6 *The Dogs d'Amour* top-10 similar artists using CF with Pearson correlation distance (*left*) and conditional probability (*right*).

We can see that the asymmetric conditional probability metric is completely biased towards popular artists, whilst Pearson similarity contains artists across the long tail, also ranging different styles (including some unexpected results, such as *Michael Jackson* or *Zero 7*). Top-10 similar artists list, obtained by conditional probability, contain some of the most representative and prototypical artists of the seed artist's main styles (that is, *glam*, *rock*, and *hard-rock*). The similarity value using conditional probability is also quite informative; 48.4% of the users who listen to *The Dogs d'Amour* also listen to *Guns n' Roses* (but not the other way around!).

3.4.2 Context-Based Filtering

As introduced in Sec. 3.3.3, context-based filtering uses cultural information to compute artist or song similarity. Context-based filtering is based on web mining techniques, or mining data from collaborative tagging (see Sec. 2.5.4).

---

[16] http://en.wikipedia.org/wiki/The_Dogs_D'Amour

3.4.2.1 An Example

Now, we present some examples from the 3-order tensor of $\langle user, artist, tag, value \rangle$ triples, using data from *last.fm*. The dataset contains 84,838 artists, and 187,551 distinct tags. We decompose the tensor, and use the artist-tag A matrix. $A_{i,j}$ contains the (*last.fm* normalised) relevance of tag j for artist i. Then, we apply Latent Semantic Analysis (LSA). LSA uses Singular Value Decomposition (SVD) to infer the hidden relationships in the data. LSA is used in Information Retrieval to compute document similarity, and also to detect term similarity (e.g. synonyms). In our case, we can consider that a document equals to an artist, and the terms that appear in the document are the artist's tags. Then, we apply SVD to reduce the matrix A to 100 factors (dimensions). After that, cosine similarity is used to derive artist similarity. Table 3.7 shows the top-10 similar artists to *The Dogs d'Amour*.

The Dogs d'Amour	Similarity$_{LSA}$
d-a-d	0.9605
Mike tramp	0.9552
Metal majesty	0.9541
Nightvision	0.9540
Bulent ortacgil — sebnem ferah	0.9540
Marty casey and lovehammers	0.9540
Hey hey jump	0.9539
Camp freddy	0.9538
Hard rocket	0.9537
Paine	0.9536

Table 3.7 *The Dogs d'Amour* top-10 similar artists using social tagging data from *last.fm*. Similarity is computed using LSA (SVD with 100 factors, and cosine distance) from the artist-tag matrix.

One problem using this approach is that the distance to the seed artist (in the 100-dimensional space) is very high, even for an artist at position top-100 in the similarity list. For instance, *The Dogs d'Amour* top-20 similar artist, *Gilby Clarke*, has a similarity value of 0.936, and the artist at top-100 (*Babylon A.D.*) has 0.868. Both artists could easily appear in the list of *The Dogs d'Amour* similar artists, but probably they will not (at least, they will not appear in the first page). Then, when presenting a list of *The Dogs d'Amour* similar artists, the user can miss some artists that are at position top-80, and that are still relevant. This happens because the semantic distance based on tags (using the 100 factors after applying SVD) is very coarse. To overcome this problem, we present in sec. 3.4.3 a hybrid approach that combines collaborative filtering and social tagging, producing more reliable results.

3.4.3 Content-Based Filtering

In the music domain, audio content-based methods rank tracks based on how similar they are according to a seed song. A music recommender system using audio content-based analysis has to compute the similarity among songs, in order to recommend music to the user. Artist similarity can also be computed, by agreggating song similarity results. There are two orthogonal ways to describe the audio content; manually or automatically.

3.4.3.1 Music Similarity Based on Manual Audio Content Description

Human-based annotation of music is very time consuming, but can be more accurate than automatic feature extraction methods. Pandora's approach is based on manual descriptions of the audio content. Pandora's web site explains their procedure[17]:

> (...) our team of thirty musician-analysts have been listening to music, one song at a time, studying and collecting literally hundreds of musical details on every song. It takes 20-30 minutes per song to capture all of the little details that give each recording its magical sound —melody, harmony, instrumentation, rhythm, vocals, lyrics ... and more— close to 400 attributes! (...)

The analysts have to annotate around 400 parameters per song, using a ten point scale [0..10] per attribute. There is a clear scalability problem; time-constraints allow people to add about 15,000 songs per month. Also, they have to deal with the variability across the analysts. Cross validation is also needed in order to assure the quality (and avoid analysts' bias) of the annotations.

Simple weighted Euclidean distance is used to find similar songs.[18] Song selection is, then, based on nearest neighbors. However, they assign specific weights to important attributes, such as genre. For artist similarity they only use specific songs, not an average of all the artist's songs. *Pandora*'s ultimate goal is to offer a mix of familiarity, diversity, and discovery.

3.4.3.2 Music Similarity Based on Automatic Audio Content Description

Early work on audio similarity is based on low-level descriptors, such as Mel Frequency Cepstral Coefficients (MFCC). These approaches aimed at deriving timbre similarity, but have also been used to take on other problems, such as genre classification. Foote proposed a music indexing system based on MFCC histograms in [73]. Audio features are usually aggregated together using mean and variance, or modelling it as a Gaussian Mixture Model (GMM). Using mean and variance for

---

[17] http://www.pandora.com/corporate/index.shtml Last accessed date: September 10th, 2008

[18] Personal communication with Pandora staff, on July 2007, while preparing the Music Recommendation Tutorial for the 2007 ISMIR conference.

the first N MFCCs, a simple (weighted) Euclidean distance, $d(X,Y)$, can be used to compute audio similarity between two songs X and Y.

Aucouturier et al. present a Gaussian mixture model (GMM) based on MFCCs [74]. Similarity measures on top of the GMM timbre model includes Kullback–Leibler (KL) divergence, and the Earth Mover's distance (EMD). KL divergence measures the relative similarity between two Gaussian distributions of data. A small divergence in the distributions means that the two songs are similar. Equation (3.1) shows a closed form symmetric approximation of the Kullback–Leibler divergence between two songs X and Y. The timbre model used is a single Gaussian with full covariance matrix.

$$d(X,Y) = Tr(\Sigma_X^{-1}\Sigma_Y) + Tr(\Sigma_Y^{-1}\Sigma_X) + Tr((\Sigma_X^{-1}\Sigma_Y^{-1})(\mu_X - \mu_Y)(\mu_X - \mu_Y)^T) - 2N_{MFCC}$$
(3.1)

where μ_X and μ_Y are MFCC means, Σ_X and Σ_Y are MFCC covariance matrices, $Tr(M)$ the trace of matrix M, and N_{MFCC} the number of MFCCs used (for instance, 13).

Earth Mover's distance (EMD) has been largely applied in the image community to retrieve similar images [75]. The EMD is defined as the minimum amount of work needed to change one (audio) signature to another. The adoption of this distance in the music field was first introduced by Logan in [76], where audio signatures are modelled with a GMM.

However, none of these two methods capture information about long-term structure elements, such as the melody, ryhthm, or harmony. To cope with this limitation, Tzanetakis extracted a set of features representing the spectrum, rhythm and harmony (chord structure) [77]. Audio features are then merged into a single vector, and are used to determine song similarity. For a complete overview on audio similarity, the reader is referred to [78]. One step further, Slaney et al. present in [79] machine learning techniques to derive a robust metric for music similarity.

Cataltepe et al. present a music recommendation system based on audio similarity [80], where user's listening history is taken into account. The hypothesis is that users give more importance to different aspects of music. These aspects can be described and classified using semantic audio features. Using this adaptative content-based recommendation scheme, as opposed to a static set of features, resulted in up to 60% of increment in the accuracy of the recommendations.

User's relevance feedback for a content-based music similarity system is presented in [81]. To reduce the burden of users to input learning data into the system, they propose a method to generate user profiles based on genre preferences, and a posterior refinement based on relevance feedback from the recommendations [82].

Once the audio has been semantically annotated (see Sec. 3.3.4), and the audio similarity among the songs has been computed, content-based filtering for a given user is rather simple. It is based on presenting songs (or artists) that "sound" similar to the user profile.

3.4.3.3 An Example

Now we present an example of artist similarity derived from automatic audio feature extraction. To compute artist similarity, we apply content-based audio analysis in an in-house music collection ($\mathcal{T}$) of 1.3 Million tracks of 30 secs. samples.

Distance between tracks, $d(x, y)$, is based on the Euclidean distance over a reduced space using Principal Component Analysis (PCA). Audio features include not only timbral features (e.g. Mel frequency cepstral coefficients), but musical descriptors related to rhythm (e.g. beats per minute, perceptual speed, binary/ternary metric), and tonality (e.g. chroma features, key and mode), among others [83]. Preliminary steps to compute the Euclidean distance are: (i) audio descriptor normalisation in the $[0, 1]$ interval, and (ii) applying PCA to reduce the audio descriptors space to 25 dimensions.

To compute artist similarity we used the most representative tracks, $\mathcal{T}_a$, of an artist a, with a maximum of 100 tracks per artist. For each track, $t_i \in \mathcal{T}_a$, we obtain the most similar tracks (excluding those from artist a).

$$sim(t_i) = \operatorname*{argmin}_{\forall t \in \mathcal{T}} \left(d(t_i, t) \right), \tag{3.2}$$

and get the artists' names, $\mathcal{A}_{sim(t_i)}$, of the t_i similar tracks. The list of (top-20) similar artists of a comes from $\mathcal{A}_{sim(t_i)}$, ranked by a combination of the artist frequency (how many songs from the artist are similar to seed track t_i), and the similarity distance (Eq. 3.3).

$$similar\_artists(a) = \bigcup \mathcal{A}_{sim(t_i)}, \forall t_i \in \mathcal{T}_a \tag{3.3}$$

Table 3.8 shows the top-20 similar artists for two seed artists, *Aerosmith*.[19] and *Alejandro Sanz*.[20] Regarding *Aerosmith*'s top-20 similar artists, most of the bands belong to the same genre, that is *classic hard rock*. Yet, some bands belong to the punk/rock style (e.g. *NOFX, MxPx, New Found Glory, Slick Shoes,* and *The Damned*). These bands could still be considered relevant to a user that has a musical taste ranging from *classic hard rock* to *punk/rock* styles. However, there are two surprising and unexpected results. These are *Die schäfer* and *Die flippers*. Both bands fall into the German folk/pop style, and their music is very different from *Aerosmith* (or any other band in the *Aerosmith*'s top-20 similar artists). Our guess is that they appear due to *Aerosmith*'s quiet pop/rock ballads. Still, these two German artists can be considered as "outliers".

Alejandro Sanz is a Spanish singer/songwriter. His music fits into latin pop, ballads, and soft rock, all merged with a flamenco touch. Even though content-based is context agnostic, some similar artists also sing in Spanish (*Gipsy Kings, Ricky Martin, Presuntos Implicados, Luis Miguel, Laura Pausini, Miguel Bosé* and *Maná*). Furthermore, most of the similar artists come from his pop songs, like *Ricky Martin,*

---

[19] For more information about the band see http://en.wikipedia.org/wiki/Aerosmith

[20] For more information about the artist see http://en.wikipedia.org/wiki/Alejandro _Sanz

Aerosmith	Similarity$_{CB}$	Alejandro Sanz	Similarity$_{CB}$
Bon Jovi	3.932	Ricky Martin	3.542
.38 Special	3.397	Jackson Browne	2.139
Guns N' Roses	3.032	Gipsy Kings	1.866
Def Leppard	2.937	Presuntos Implicados	1.781
Ozzy Osbourne	2.795	Emmylou Harris	1.723
Helloween	2.454	Luis Miguel	1.668
Kiss	2.378	Laura Pausini	1.529
Bryan Adams	2.180	Ry Cooder	1.479
Poison	2.088	Harry Chapin	1.370
The Damned	2.044	Dwight Yoakam	1.332
Tesla	2.030	Nek	1.331
Die Schäfer	1.963	Miguel Bosé	1.298
Mötley Crüe	1.949	Maná	1.241
Nofx	1.807	The Doobie Brothers	1.235
MXPX	1.733	Uncle Kracker	1.217
New Found Glory	1.718	Seal	1.184
Slick Shoes	1.677	Anika Moa	1.174
Die Flippers	1.662	Graham Central Station	1.158
Uriah Heep	1.659	The Imperials	1.157
Alice Cooper	1.608	The Corrs	1.152

Table 3.8 Similar artists for *Aerosmith* (left column) and *Alejandro Sanz* (right column).

Presuntos Implicados, Nek, Seal, Maná, Miguel Bosé and *The Corrs*. His flamenco and acoustic facets are also present in the *Gipsy Kings* band. *Luis Miguel* appears in the list because of *Alejandro Sanz*'s quiet ballads. The rest of the artists fall into the broad range of singer/songwriter, folk and Americana styles, and includes: *Jackson Browne, Emmy Lou Harris, Ry Cooder, Dwight, Uncle Kracker* and *Harry Chapin*. In this case, similarity with *Alejandro Sanz* is more arguably. Also, a few similar artists are female singers (*Anika Moa, The Corrs, Presuntos Implicados, Emmylou Harris*, and *Laura Pausini*). In these cases, music similiarty and production artifacts probably predominate over melody and voice. Finally, there are some strange and incomprehensible artists, such as *Graham Central Station* (a long tail band, playing a mix of funk, soul, and rhythm and blues), and *The Imperials* (also a long tail band, that plays doo-wop and gospel music). Without any explanation or transparency about these recommendations, a user will probably perceive some of the similar artists as non-relevant.

3.4.4 Hybrid Methods

The combination of different approaches allows a system to minimise the issues that a solely method can have. One way to combine different recommendation methods is the cascade approach (see Sec. 2.5.5). Cascade is a step by step process. One technique is applied first, obtaining a ranked list of items. Then, a second technique refines or re-rank the results obtained in the first step.

For example, to compute artist similarity a system can first apply CF, and then reorder or combine the results according to the semantic distance from social tagging (LSA). Another option is first apply CF as well as the semantic distance from social tagging (LSA), and combine these results. After that, apply content-based audio similarity to rerank the similar artists.

3.4.4.1 An example

Table 3.9 shows *The Dogs d'Amour* similar artists using a cascade hybrid method. First, *The Dogs d'Amour* top-100 similar artists are computed using CF, with Pearson correlation distance. In a second step, for each artist in this top-100 list we compute LSA—using SVD with 100 factors—and cosine similarity from the social tagging data, between the actual artist and the seed artist (*The Dogs d'Amour*). After that, we combine the results from Pearson CF with the results obtained in this second step. We use a linear combination function setting $\alpha = 0.5$:

$$sim(a_i, a_j)_{Hybrid} = (1 - \alpha) \cdot sim(a_i, a_j)_{CF,Pearson} + \alpha \cdot sim(a_i, a_j)_{Context,LSA} \quad (3.4)$$

This way we can improve the original CF results, and also the results obtained solely from social tagging. Indeed, the Pearson CF approach returned some strange and non-relevant results, such as *Michael Jackson* or *Zero 7* (see Table 3.9, left). After reordering the results using social tagging data, both artists disappear from the top-10 (hybrid) list of similar artists. Also, some artists that were not in the CF top-10 appear in the final set of similar artists (Table 3.9, right), due to the linear combination of the two approaches (Pearson CF and LSA from tags).

In this case, the cascade chain method makes sense. The first results are obtained taking into account the music users listen to; "people who listen to *The Dogs d'Amour* also listen to X". Then, the second step promotes those artists X that are closer, in the semantic community annotation space, to the seed artist.[21]

3.4.4.2 Related Work

Related work in hybrid music recommendation is presented in Yoshii et al. [84, 85]. The origins of their work can be found in [86], where they present a hybrid music recommender system based on a probabilistic generative model, named three-way aspect model [87]. The model explains the generative process for the observed data by introducing a set of latent variables. Their system integrates both explicit collaborative filtering and audio content-based features. Collaborative filtering contains the users' ratings for the songs, and it is based on a [0..2] scale. A zero means

---

[21] After some inspection, and according to the author's knowledge of *The Dogs d'Amour* band, the hybrid approach produces much better results than both LSA from social tagging and Pearson CF alone.

The Dogs d'Amour	Similarity$_{Pearson}$	The Dogs d'Amour	Similarity$_{Hybrid}$
Los Fabulosos Cadillacs	0.806	Electric Boys	0.868
Electric Boys	0.788	Lillian Axe	0.826
Lillian Axe	0.784	Ginger	0.752
Michael Jackson	0.750	Enuff z'nuff	0.732
Ginger	0.723	Michael Monroe	0.724
The Decemberists	0.699	Hardcore Superstar	0.692
The Byrds	0.667	Faster Pussycat	0.691
Zero 7	0.661	Firehouse	0.690
Rancid	0.642	Nashville Pussy	0.677
The Sonics	0.629	The Wildhearts	0.651

Table 3.9 *The Dogs d'Amour* top-10 similar artists using CF with Pearson correlation distance (left), and (right) a hybrid version using only the top-100 similar artists from CF, and reordering the artists using LSA and cosine distance from social tagging.

that the user does not like the song, 1 means indifference, and 2 that a user likes the song. Content-based audio features include a Gaussian Mixture Model using the 13 coefficients from MFCC. The authors improve the efficiency and scalability of the previous approach, using incremental learning in [85].

Tiemann et al. investigate ensemble learning methods for hybrid music recommender algorithms in [88]. This approach combines social and content-based methods, where each one produces a weak learner. Then, using a combination rule, it unifies the output of the weak learners. The results suggests that the hybrid approach reduces the mean absolute prediction error, compared to the weak learners used solely.

3.5 Summary

This chapter has presented the main actors in music recommendation; user profiling and the representation of musical items, as well as the existing methods to recommend music assets given a user profile.

User preferences depends on the type of listener, and her level of engagement with the music. Furthermore, music perception is very subjective, and it is influenced by the context. In this sense, user profile representation is an important aspect. We have presented three different notations: UMIRL, MPEG-7 based, and Friend of a Friend. The former is one of the first attempts in this field. The UMIRL language is not formal enough, but a proposal that contains some interesting ideas. User preferences in MPEG-7 is the first big and serious attempt to formalise user modelling, related with the multimedia content. The main problem of this approach is that the MPEG-7 standard is too complex and verbose. It is not straight forward to generate user profiles following the notation proposed by the standard. The last proposal, Friend of a Friend profiles, is based on the Semantic Web initiative. It is the most flexible approach. As it is based on the Semantic Web premises, Friend of a Friend profiles can embed different ontologies, so it is extensible, and has richer semantics than the other two approaches.

In music recommedation, item-based similarity is the most common way to compute and predict the recommendations. Item profile representation, then, is the first step to compute item similarity, in order to provide music recommendations to a user. We describe the representation and modelling of music items via the Music Information Plane. MIP defines the different levels of complexity and abstraction of the music descriptions. Based on the MIP approach, we present three complementary ways to describe artists and songs; using editorial, cultural, and acoustic information. Similarity measures—based on the editorial, cultural, and acoustic information—are also introduced. Then, for each recommendation method, we present the resulting list of top-20 similar artists using *The Dogs d'Amour* rock band as seed artist. An informal evaluation shows that the hybrid approach, using a mix of collaborative filtering and social tagging, produces the best results.

3.5.1 Links with the Following Chapters

An important remaining task is the formal evaluation of music (and user) similarity, as this is the basis to provide music recommendations. This evaluation is presented in Chap. 5, that presents the metrics, and Chap. 6, that contains the actual evaluation of real, and big datasets. Also, user's perceived quality of the recommendations is very important. We present, in Chap. 7, an experiment done with 288 subjects, that analyses the effects of providing novel and relevant music recommendations to users. Still, before going further into the evaluation process, we present in Chap. 4 the Long Tail phenomenon, and its effects in the music domain.

References

1. C. Baccigalupo and E. Plaza, "Poolcasting: a social web radio architecture for group customisation," in *Proceedings of the 3rd International Conference on Automated Production of Cross Media Content for Multi-Channel Distribution AXMEDIS*, pp. 115–122, IEEE Computer Society, 2007.
2. A. Crossen, J. Budzik, and K. J. Hammond, "Flytrap: Intelligent group music recommendation," in *Proceedings of the 7th International Conference on Intelligent User Interfaces*, (New York, NY, USA), pp. 184–185, ACM, 2002.
3. S. J. Cunningham, D. Bainbridge, and A. Falconer, "More of an art than a science: Supporting the creation of playlists and mixes," in *Proceedings of 7th International Conference on Music Information Retrieval*, (Victoria, Canada), pp. 240–245, 2006.
4. T.W. Leong, F. Vetere, and S. Howard, "The serendipity shuffle," in *Proceedings of 19th Conference of the Computer-Human Interaction Special Interest Group*, (Narrabundah, Australia), pp. 1–4, 2005.
5. A. Uitdenbogerd and R. van Schnydel, "A review of factors affecting music recommender success," in *Proceedings of 3rd International Conference on Music Information Retrieval*, (Paris, France), 2002.

6. D. Jennings, *Net, Blogs and Rock 'n' Roll: How Digital Discovery Works and What it Means for Consumers*. Boston, MA: Nicholas Brealey Publishing, 2007.

7. M. Lesaffre, M. Leman, and J.-P. Martens, "A user-oriented approach to music information retrieval," in *Content-Based Retrieval*, Dagstuhl Seminar Proceedings, 2006.

8. F. Vignoli and S. Pauws, "A music retrieval system based on user driven similarity and its evaluation," in *Proceedings of the 6th International Conference on Music Information Retrieval*, (London, UK), pp. 272–279, 2005.

9. D. N. Sotiropoulos, A. S. Lampropoulos, and G. A. Tsihrintzis, "Evaluation of modeling music similarity perception via feature subset selection," in *User Modeling*, vol. 4511 of *Lecture Notes in Computer Science*, (Berlin, Heidelberg), pp. 288–297, Springer, 2007.

10. V. Sandvold, T. Aussenac, O. Celma, and P. Herrera, "Good vibrations: Music discovery through personal musical concepts," in *Proceedings of 7th International Conference on Music Information Retrieval*, (Victoria, Canada), 2006.

11. P. Kazienko and K. Musial, "Recommendation framework for online social networks," in *Advances in Web Intelligence and Data Mining*, vol. 23 of *Studies in Computational Intelligence*, pp. 111–120, Springer, 2006.

12. S. Baumann, B. Jung, A. Bassoli, and M. Wisniowski, "Bluetuna: Let your neighbour know what music you like," in *CHI – extended abstracts on Human factors in computing systems*, (New York, NY, USA), pp. 1941–1946, ACM, 2007.

13. C. S. Firan, W. Nejdl, and R. Paiu, "The benefit of using tag-based profiles," in *Proceedings of the 2007 Latin American Web Conference (LA-WEB)*, (Washington, DC, USA), pp. 32–41, IEEE Computer Society, 2007.

14. E. Perik, B. de Ruyter, P. Markopoulos, and B. Eggen, "The sensitivities of user profile information in music recommender systems," in *Proceedings of Private, Security, Trust*, (New Brunswick, Canada), 2004.

15. W. Chai and B. Vercoe, "Using user models in music information retrieval systems," in *Proceedings of 1st International Conference on Music Information Retrieval*, (Berlin), 2000.

16. B. S. Manjunath, P. Salembier, and T. Sikora, *Introduction to MPEG 7: Multimedia Content Description Language*. Ed. Wiley, 2002.

17. C. Tsinaraki and S. Christodoulakis, "Semantic user preference descriptions inMPEG-7/21," in *Hellenic Data Management Symposium*, (Athens, Greece), 2005.

18. Y. Raimond, S. A. Abdallah, M. Sandler, and F. Giasson, "The music ontology," in *Proceedings of the 8th International Conference on Music Information Retrieval*, (Vienna, Austria), 2007.

19. F. Pachet, *Knowledge Management and Musical Metadata*. Idea Group, 2005.

20. B. Whitman and S. Lawrence, "Inferring descriptions and similarity for music from community metadata," in *Proceedings of International Computer Music Conference*, (Goteborg, Sweden), 2002.

21. M. Schedl, P. Knees, T. Pohle, and G. Widmer, "Towards an automatically generated music information system via web content mining," in *Proceedings of the 30th European Conference on Information Retrieval (ECIR'08)*, (Glasgow, Scotland), 2008.

22. P. Knees, M. Schedl, and T. Pohle, "A deeper look into web-based classification of music artists," in *Proceedings of 2nd Workshop on Learning the Semantics of Audio Signals*, (Paris, France), 2008.

23. B. Whitman, "Semantic rank reduction of music audio," in *Proceedings of the Workshop on Applications of Signal Processing to Audio and Acoustics (WASPAA)*, (New paltz, NY, USA), pp. 135–138, 2003.

24. S. Baumann and O. Hummel, "Enhancing music recommendation algorithms using cultural metadata," *Journal of New Music Research*, vol. 34, no. 2, 2005.

25. G. Geleijnse and J. Korst, "Web-based artist categorization," in *Proceedings of the 7th International Conference on Music Information Retrieval*, (Victoria, Canada), pp. 266–271, 2006.

26. M. Zadel and I. Fujinaga, "Web services for music information retrieval," in *Proceedings of 5th International Conference on Music Information Retrieval*, (Barcelona, Spain), 2004.

27. M. Schedl, P. Knees, and G. Widmer, "A web-based approach to assessing artist similarity using co-occurrences," in *Proceedings of 4th International Workshop on Content-Based Multimedia Indexing*, (Riga, Latvia), 2005.

28. M. Schedl, P. Knees, and G. Widmer, "Improving prototypical artist detection by penalizing exorbitant popularity," in *Proceedings of 3rd International Symposium on Computer Music Modeling and Retrieval*, (Pisa, Italy), pp. 196–200, 2005.

29. T. Pohle, P. Knees, M. Schedl, and G. Widmer, "Building an interactive next-generation artist recommender based on automatically derived high-level concepts," in *Proceedings of the 5th International Workshop on Content-Based Multimedia Indexing*, (Bordeaux, France), 2007.

30. C. Baccigalupo, J. Donaldson, and E. Plaza, "Uncovering affinity of artists to multiple genres from social behaviour data," in *Proceedings of the 9th Conference on Music Information Retrieval*, (Philadelphia, Pennsylvania USA), pp. 275–280, 2008.

31. F. Pachet, G. Westermann, and D. Laigre, "Musical data mining for electronic music distribution," in *Proceedings of 1st International Conference on Web Delivering of Music*, 2001.

32. D. Ellis, B. Whitman, A.Berenzweig, and S.Lawrence, "The quest for ground truth in musical artist similarity," in *Proceedings of 3rd International Symposium on Music Information Retrieval*, (Paris), pp. 170–177, 2002.

33. G. Geleijnse, M. Schedl, and P. Knees, "The quest for ground truth in musical artist tagging in the social web era," in *Proceedings of the 8th International Conference on Music Information Retrieval*, (Vienna, Austria), 2007.

34. A. Berenzweig, B. Logan, D. Ellis, and B. Whitman, "A large-scale evalutation of acoustic and subjective music similarity measures," in *Proceedings of 4th International Symposium on Music Information Retrieval*, (Baltimore, MD), 2003.

35. D. Turnbull, L. Barrington, and G. Lanckriet, "Five approaches to collecting tags for music," in *Proceedings of the 9th International Conference on Music Information Retrieval*, (Philadelphia, PA), pp. 225–230, 2008.

36. G. Koutrika, F. A. Effendi, Z. Gyöngyi, P. Heymann, and H. Garcia-Molina, "Combating spam in tagging systems," in *Proceedings of the 3rd international workshop on Adversarial information retrieval on the web*, (New York, NY), pp. 57–64, ACM, 2007.

37. L. R. Rabiner and B. H. Juang, *Fundamentals of Speech Recognition*. Englewood Cliffs, NJ: Prentice-Hall, 1993.

38. J.-J. Aucouturier and F. Pachet, "A scale-free distribution of false positives for a large class of audio similarity measures," *Pattern Recognition*, vol. 41, no. 1, pp. 272–284, 2008.

39. J.-J. Aucouturier and F. Pachet, "Improving timbre similarity: How high's the sky," *Journal of Negative Results in Speech and Audio Science*, vol. 1, no. 1, 2004.

40. P. Herrera, V. Sandvold, and F. Gouyon, "Percussion-related semantic descriptors of music audio files," in *Proceedings of 25th International AES Conference*, (London, UK), 2004.

41. K. Yoshii, M. Goto, and H. G. Okuno, "Automatic drum sound description for real-world music using template adaptation and matching methods," in *Proceedings of 5th International Conference on Music Information Retrieval*, (Barcelona, Spain), 2004.

42. N. Chetry, M. Davies, and M. Sandler, "Musical instrument identification using LSF and Kmeans," in *Proceedings of the 118th Convention of the AES*, (Barcelona, Spain), 2005.

43. F. Gouyon and S. Dixon, "A review of automatic rhythm description systems," *Computer Music Journal*, vol. 29, pp. 34–54, 2005.

44. P. Bello and M. Sandler, "Phase-based note onset detection for music signals," in *Proceedings of IEEE ICASSP*, (Hong Kong, China), 2003.

45. J. P. Bello, C. Duxbury, M. E. Davies, and M. B. Sandler, "On the use of phase and energy for musical onset detection in the complex domain," in *IEEE Signal Processing Letters*, pp. 533–556, 2004.

46. M. E. P. Davies and M. D. Plumbley, "Causal tempo tracking of audio," in *Proceedings of 5th International Conference on Music Information Retrieval*, (Barcelona, Spain), 2004.

47. F. Gouyon and S. Dixon, "Dance music classification: A tempo-based approach," in *Proceedings of 5th International Conference on Music Information Retrieval*, (Barcelona, Spain), 2004.

48. S. Dixon, F. Gouyon, and G. Widmer, "Towards characterization of music via rhythmic patterns," in *Proceedings of 5th International Conference on Music Information Retrieval*, (Barcelona, Spain), 2004.

49. R. Dannenberg, "Toward automated holistic beat tracking, music analysis, and understanding," in *Proceedings of 6th International Conference on Music Information Retrieval*, (London, UK), 2005.

50. J. Pickens, J. P. Bello, G. Monti, T. Crawford, M. Dovey, M. Sandler, and D. Byrd, "Polyphonic score retrieval using polyphonic audio queries: A harmonic modelling approach," in *Proceedings of 3rd International Conference on Music Information Retrieval*, pp. 140–149, 2002.

51. E. Gómez, "Tonal description of polyphonic audio for music content processing," *INFORMS Journal on Computing, Special Cluster on Computation in Music*, vol. 18, no. 3, 2006.

52. E. Gómez and P. Herrera, "Estimating the tonality of polyphonic audio files: Cognitive versus machine learning modelling strategies," *Proceedings of 5th International Conference on Music Information Retrieval*, (Barcelona, Spain), 2004.

53. C. A. Harte and M. Sandler, "Automatic chord identification using a quantised chromagram," in *Proceedings of the 118th Convention of the AES*, (Barcelona, Spain), 2005.

54. P. Bello and J. Pickens, "A robust mid-level representation for harmonic content in music signals," in *Proceedings of 6th International Conference on Music Information Retrieval*, (London, UK), 2005.

55. E. Gómez, *Tonal Description of Music Audio Signals*. PhD thesis, Universitat Pompeu Febra, Barcelona, Spain 2006.

56. B. Ong and P. Herrera, "Semantic segmentation of music audio contents," in *Proceedings of International Computer Music Conference*, (Barcelona, Spain), 2005.

57. A. Zils and F. Pachet, "Extracting automatically the perceived intensity of music titles," in *Proceedings of the 6th International Conference on Digital Audio Effects*, (London, UK), 2003.

58. V. Sandvold and P. Herrera, "Towards a semantic descriptor of subjective intensity in music," in *Proceedings of 5th International Conference on Music Information Retrieval*, (Barcelona, Spain), 2004.

59. F. Fabbri, "Browsing music spaces: Categories and the musical mind," in *Proceedings of the IASPM Conference*, 1999.

60. D. Brackett, *Intepreting Popular Music*. New York, NY: Canbridge University Press, 1995.

61. C. Mckay and I. Fujinaga, "Musical genre classification: Is it worth pursuing and how can it be improved?," in *Proceedings of the 7th International Conference on Music Information Retrieval*, (Victoria, Canada), 2006.

62. G. Tzanetakis and P. Cook, "Musical genre classification of audio signals," *IEEE Transactions on Speech and Audio Processing*, vol. 10, no. 5, pp. 293–302, 2002.

63. E. Guaus, *Audio content processing for automatic music genre classification: descriptors, databases, and classifiers*. PhD thesis, 2009.

64. P. N. Juslin and P. Laukka, "Expression, perception, and induction of musical emotions: A review and a questionnaire study of everyday listening," *Journal of New Music Research*, vol. 22, no. 1, pp. 217–238, 2004.

65. P. N. Juslin and J. A. Sloboda, *Music and Emotion: Theory and Research*. Oxford: Oxford University Press, 2001.

66. Y. Feng, Y. Zhuang, and Y. Pan, "Music information retrieval by detecting mood via computational media aesthetics," in *WI '03: Proceedings of the 2003 IEEE/WIC International Conference on Web Intelligence*, (Washington, DC, USA), p. 235, IEEE Computer Society, 2003.

67. D. Z. H. Lu, L Liu, "Automatic mood detection and tracking of music audio signals," *IEEE Transactions on Audio, Speech and Language Processing*, vol. 14, pp. 5–18, 2006.

68. C. Laurier, O. Lartillot, T. Eerola, and P. Toiviainen, "Exploring relationships between audio features and emotion in music," in *Conference of European Society for the Cognitive Sciences of Music*, (Jyväskylä, Finland), 2009.

69. K. Bischoff, C. Firan, R. Paiu, W. Nejdl, C. Laurier, and M. Sordo, "Music mood and theme classification a hybrid approach," in *Proceedings of the 10th Conference on Music Information Retrieval*, (Kobe, Japan), 2009.

70. U. Shardanand, "Social information filtering for music recommendation," Master's thesis, Massachussets Institute of Technology, September 1994.
71. M. Anderson, M. Ball, H. Boley, S. Greene, N. Howse, D. Lemire, and S. McGrath, "Racofi: A rule-applying collaborative filtering system," in *Proceedings of the Collaboration Agents Workshop*, IEEE/WIC, (Halifax, Canada), 2003.
72. D. Lemire and A. Maclachlan, "Slope one predictors for online rating-based collaborative filtering," in *Proceedings of SIAM Data Mining*, (Newport Beach, CA), 2005.
73. J. Foote, "Content-based retrieval of music and audio," *Multimedia Storage and Archiving Systems II. Proceedings of SPIE*, pp. 138–147, 1997.
74. J.-J. Aucouturier and F. Pachet, "Music similarity measures: What's the use?," in *Proceedings of 3rd International Conference on Music Information Retrieval*, (Paris, France), pp. 157–163, 2002.
75. C. T. Y. Rubner and L. Guibas, "The earth mover's distance as a metric for image retrieval," tech. rep., Stanford University, 1998.
76. B. Logan and A. Salomon, "A music similarity function based on signal analysis," *IEEE International Conference on Multimedia and Expo, 2001. ICME 2001*, pp. 745–748, 2001.
77. G. Tzanetakis, *Manipulation, Analysis and Retrieval Systems for Audio Signals*. PhD thesis, 2002.
78. E. Pampalk, *Computational Models of Music Similarity and their Application to Music Information Retrieval*. PhD thesis, 2006.
79. M. Slaney, K. Weinberger, and W. White, "Learning a metric for music similarity," in *Proceedings of the 9th Conference on Music Information Retrieval*, (Philadelphia, Pennsylvania, USA), pp. 313–318, 2008.
80. B. Cataltepe, Z. Altinel, "Music recommendation based on adaptive feature and user grouping," in *Proceedings of the 22nd International International Symposium on Computer and Information Sciences*, (Ankara, Turkey), 2007.
81. K. Hoashi, K. Matsumoto, and N. Inoue, "Personalization of user profiles for content-based music retrieval based on relevance feedback," in *Proceedings of eleventh ACM international conference on Multimedia*, (New York, NY, USA), pp. 110–119, ACM Press, 2003.
82. J. J. Rocchio, "Relevance feedback in information retrieval," in *The SMART Retrieval System: Experiments in Automatic Document Processing* (G. Salton, ed.), Prentice-Hall Series in Automatic Computation, ch. 14, pp. 313–323, Englewood Cliffs, NJ: Prentice-Hall, 1971.
83. P. Cano, M. Koppenberger, and N. Wack, "An industrial-strength content-based music recommendation system," in *Proceedings of 28th International ACM SIGIR Conference*, (Salvador, Brazil), 2005.
84. K. Yoshii, M. Goto, K. Komatani, T. Ogata, and H. G. Okuno, "An efficient hybrid music recommender system using an incrementally trainable probabilistic generative model," *IEEE Transaction on Audio Speech and Language Processing*, vol. 16, no. 2, pp. 435–447, 2008.
85. K. Yoshii, M. Goto, K. Komatani, T. Ogata, and H. G. Okuno, "Improving efficiency and scalability of model-based music recommender system based on incremental training," in *Proceedings of 8th International Conference on Music Information Retrieval*, (Vienna, Austria), 2007.
86. K. Yoshii, M. Goto, K. Komatani, T. Ogata, and H. G. Okuno, "Hybrid collaborative and content-based music recommendation using probabilistic model with latent user preferences," in *Proceedings of 7th International Conference on Music Information Retrieval*, (Victoria, Canada), pp. 296–301, 2006.
87. A. Popescul, L. Ungar, D. Pennock, and S. Lawrence, "Probabilistic models for unified collaborative and content-based recommendation in sparse-data environments," in *17th Conference on Uncertainty in Artificial Intelligence*, (Seattle, Washington), pp. 437–444, 2001.
88. M. Tiemann and S. Pauws, "Towards ensemble learning for hybrid music recommendation," in *Proceedings of 8th International Conference on Music Information Retrieval*, (Vienna, Austria), 2007.

Chapter 4
The Long Tail in Recommender Systems

4.1 Introduction

The Long Tail is composed of a small number of popular items, the well-known *hits*, and the rest are located in the heavy tail, those not sell *that well*. The Long Tail offers the possibility to explore and discover—using automatic tools; such as recommenders or personalised filters—vast amounts of data. Until now, the world was ruled by the *Hit or Miss* categorisation, due in part to the shelf space limitation of the brick-and-mortar stores. A world where a music band could only succeed selling millions of albums, and touring worldwide.

Nowadays, we are moving towards the *Hit vs. Niche* paradigm, where there is a large enough availability of choice to satisfy even the most *Progressive–obscure–Spanish–metal* fan. The problem, though, is to filter and present the *right* artists to the user, according to her musical taste.

Chris Anderson introduces in his book, "The Long Tail" [1], a couple of important conditions to exploit the content available in niche markets. These are: (i) make everything available, and (ii) help me find it. It seems that the former condition is already fulfilled; the distribution and inventory costs are nearly negligible. Yet, to satisfy the latter we need recommender systems that exploit the *from hits to niches* paradigm. The main question, though, is whether current recommendation techniques are ready to assist us in this discovery task, providing recommendations of the *hidden gems* in the Long Tail.

In fact, recommenders that appropriately discount popularity may increase total sales, as well as potentially increase the margins by suggesting more novel, or less known, products [2]. Tucker et al. develop a theoretical model which shows how the existence of popular items can, in fact, benefit the perceived quality of niche products [3]. As these niche items are less likely to attract customers, the ones they attract perceive the products as higher quality than the mainstream ones. The authors' findings contribute to the understanding that popularity affects the long tail of e-Commerce. Even though web 2.0 tools based on the user's history of purchases promote the popular goods, their results suggest that mainstreamness benefits the

Ò. Celma, *Music Recommendation and Discovery*,
DOI 10.1007/978-3-642-13287-2_4, © Springer-Verlag Berlin Heidelberg 2010

perceived quality of niche products. Again, the big problem is to develop filters and tools that allow users to find and discover these niche products.

4.1.1 Pre- and post-filters

In the brick-and-mortar era, the market pre-filtered those products with lower probability of being bought by people. The main problem was the limited physical space to store the goods. Nowadays, with the unlimited shelf space, there is no need to pre-filter any product [1]. Instead, what users need are post-filters to make the products available and visible, and get personalised recommendations, according to their interests. Still, when publishers or producers pre-filter the content they also contribute to cultural production. E.g. many books or albums would be a lot worse without their editors and producers.

One should assume that there are some extremely poor quality products along the Long Tail. These products do not need to be removed by the gatekeepers anymore, but can remain in the Long Tail forever. The advisors are the ones in charge of not recommending low quality goods. In this sense, [4] proved that increasing the strength of social influence increased both inequality and unpredictability of success. As a consequence, popularity was only partly determined by quality. In fact, the quality of a work cannot be assessed in isolation, because our experience is so tied up with other people's experience of that work. Therefore, one can find items to match anyone's taste along the Long Tail. It is the job of the post-filters to ease the task of finding them.

4.2 The Music Long Tail

As already mentioned in Chap. 1, the "State of the Industry" report [5] presents some insights about the long tail in music consumption. For instance, 844 million digital tracks were sold in 2007, but only 1% of all digital tracks—the head part of the curve—accounted for 80% of all track sales. Also, 1,000 albums accounted for 50% of all album sales, and 450,344 of the 570,000 albums sold were purchased less than 100 times. Music consumption is biased towards a few mainstream artists. Ideally, by providing personalised filters and discovery tools to the listeners, music consumption would be diversified.

4.2.1 The Long Tail of Sales Versus the Long Tail of Plays

When computing a Long Tail distribution, one should define how to measure the popularity of the items. In the music domain, this can be achieved using the total number of sales or the total number of plays. On the one hand, the total number of sales denote the current trends in music consumption. On the other hand, the total number of playcounts tell us what people listen to, independently of the release year of the album (or song).

In terms of coverage, total playcounts is more useful, as it can represent a larger number of artists. An artist does not need to have an album released, but a *Myspace*-like page, which includes the playcounts for each song. Gathering information about the number of plays is easier than collecting the albums an artist has sold. Usually, the number of sales are shown in absolute values, aggregating all the information, and these numbers are used to compare the evolution of music consumption over the years. The total number of plays give us more accurate information, as it describes what people listen to. Thus, we will define the Long Tail in music using the total playcounts per artist.

As an example, Table 4.1 shows the overall most played artists at *last.fm* in July, 2007. These results come from more than 20 million registered users. Although the list of top-10 artists are biased towards this set of users, it still represents the listening habits of a large amount of people. In contrast, Table 4.2 shows the top-10

```
 1. The Beatles             (50,422,827)
 2. Radiohead               (40,762,895)
 3. System of a Down        (37,688,012)
 4. Red Hot Chili Peppers   (37,564,100)
 5. Muse                    (30,548,064)
 6. Death Cab for Cutie     (29,335,085)
 7. Pink Floyd              (28,081,366)
 8. Coldplay                (27,120,352)
 9. Nine Inch Nails         (24,095,408)
10. Blink 182               (23,330,402)
```

Table 4.1 Top-10 popular artists in *last.fm* according to the total number of plays (last column). Data gathered during July, 2007.

artists in 2006 based on total digital track sales (last column) according to Nielsen Soundscan 2006 report [6]. The second column (values in parenthesis) shows the corresponding *last.fm* artist rank. There is not a clear correlation between the two lists, and only one artist (*Red Hot Chili Peppers*) appears in both top-10 lists.

Furthermore, Table 4.3 shows the top-10 selling artists in 2006 based on total album sales (last column), again according to the Nielsen 2006 report. In this case, classic artists such as *Johnny Cash* (top-2) or *The Beatles* (top-5) appear. This reflects the type of users that still buy CDs. Regarding *Carrie Underwood*, she is an American country pop music singer who became famous after winning the fourth season of *American Idol* (2005). *Carrie Underwood* album, released in late 2005, became the fastest selling debut Country album. *Keith Urban*, *Tim McGraw* and *Rascal Flatts* are American country/pop songwriters with a leading male singer. In all these cases, they are not so popular in the *last.fm* community.

All in all, only *The Beatles* (in Table 4.3), and *Red Hot Chili Peppers* (in Table 4.2) appear in the top-10 *last.fm* chart (see Table 4.1). It is worth noting that in 2006 *The Beatles* music collection was not (legally) available for purchase in digital

```
 1. (912) Rascal Flatts          (3,792,277)
 2. (175) Nickelback             (3,715,579)
 3. (205) Fray                   (3,625,140)
 4. (154) All-American Rejects   (3,362,528)
 5. (119) Justin Timberlake      (3,290,523)
 6. (742) Pussycat Dolls         (3,277,709)
 7.   (4) Red Hot Chili Peppers  (3,254,306)
 8.  (92) Nelly Furtado          (3,052,457)
 9.  (69) Eminem                 (2,950,113)
10. (681) Sean Paul              (2,764,505)
```

Table 4.2 Top-10 artists in 2006 based on total digital track sales (last column) according to Nielsen report. The second column (values in parenthesis) shows the corresponding *last.fm* artist rank.

```
 1.  (912) Rascal Flatts         (4,970,640)
 2.   (70) Johnny Cash           (4,826,320)
 3.  (175) Nickelback            (3,160,025)
 4. (1514) Carrie Underwood      (3,016,123)
 5.    (1) The Beatles           (2,812,720)
 6. (1568) Tim McGraw            (2,657,675)
 7. (2390) Andrea Bocelli        (2,524,681)
 8. (1575) Mary J. Blige         (2,485,897)
 9. (1606) Keith Urban           (2,442,577)
10.  (119) Justin Timberlake     (2,437,763)
```

Table 4.3 Top-10 selling artists in 2006 (based on total album sales, last column) according to Nielsen report. The second column (values in parenthesis) shows the corresponding *last.fm* artist rank.

form. On the other hand, *last.fm* listening habits denote what people listen to, and that does not necessarily correlate with the best sellers. For instance, classic bands such as *Pink Floyd*, *Led Zeppelin* (at top-15), *Tool* (top-16) or *Nirvana* (top-18) did not release any new album during 2006, but still they are in the top-20 (at mid-2007). From this informal analysis we conclude that popularity is a nebulous concept that can be viewed in different ways.

From now on, we characterise music popularity using the total playcounts of an artist, keeping in mind that the data is not correlated with the actual number of sales, and also that the data will be biased towards the subset of users that are taken into account (in our case, the entire *last.fm* community).

4.2.2 Collecting Playcounts for the Music Long Tail

In the music field, total artist playcounts allow us to determine artist popularity. There are at least two different ways to collect artists' plays from the web. The first one is using *last.fm* data, and the second one is using the data from *Myspace*. In principle, one should expect a clear correlation among both datasets. That is, if an artist has a lot of plays in one system then the same should happen in the other one.

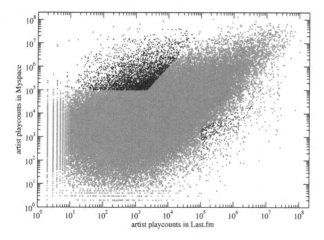

Fig. 4.1 Correlation between *last.fm* and *Myspace* artist playcounts. Data gathered during January, 2008.

However, each system measures different listening habits. On the one hand, *last.fm* monitors what users listen to in virtually any device, whereas *Myspace* only tracks the number of times a song has been played in their embedded Flash player. On the other hand, *Myspace* data can track the number of plays for those artists that have not released any album, but a list of songs (or demos) that are available on the *Myspace* artist profile. In this case, it is very unlikely to gather this data from *last.fm* because the only available source to listen to the songs is via *Myspace* (specially if the artist forbids users to download the songs from *Myspace*). For example, the artist *Thomas Aussenac* has (on October 21st, 2008) 12,486 plays in *Myspace*[1] but only 63 in *last.fm*.[2] Therefore, sometimes (e.g. head and mid artists) both systems can provide similar listening habits results, whilst in other cases they track and measure different trends. Some plausible reasons about these differences could be due to the demographics and locale of both users and artists in the two systems.

Figure 4.1 depicts the total playcounts for an artist in *last.fm* versus the total playcounts in *Myspace* (data gathered during January, 2008). That is, given the play-counts of an artist in *last.fm*, it plots its total plays in *Myspace*. We remark two inter-esting areas; upper left and bottom right. These areas are the ones with those artists whose playcounts are clearly uncorrelated between the two datasets. For instance, the upper left area shows the artists that have lots of plays in *Myspace*, but just a few in *last.fm*. The formula used to select the artists in this area is (it is analogous for the *last.fm* versus *Myspace*—in the bottom right area):

$$Plays_{Myspace} > 10^5 \wedge \frac{log(Plays_{Myspace})}{log(Plays_{Last.fm})} \geq 1.5 \qquad (4.1)$$

[1] http://www.myspace.com/thomasaussenac

[2] http://www.last.fm/music/thomas+aussenac

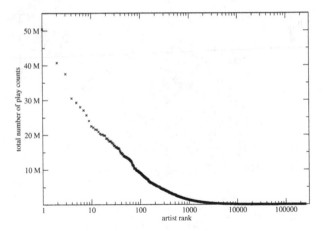

Fig. 4.2 The music Long Tail effect. A log-linear plot depicting the total number of plays per artist. Data gathered during July, 2007, for a list of 260,525 artists.

That is, artists that have more than 100,000 plays in *Myspace*, but much less in *last.fm*. In this case, we could consider that some of these artists are well-known in the *Myspace* area, having lots of fans that support them, but the artist still has no effect outside *Myspace*. Maybe this type of artists can reach a broader popularity after releasing an album. For instance, *Michael Imhof*,[3] a German *house* and *r&b* artist, has more than 200,000 playcounts in *Myspace*, but only 2 in *last.fm*. A more extreme example is *Curtis Young*[4] (aka *Hood Surgeon*), the son of legendary hip-hop producer *Dr. Dre*, who has 13,814,586 plays in *Myspace* but less than 20,000 in *last.fm*. It is worth mentioning that there are some services[5] that allow a *Myspace* artist to automatically increase their total playcounts, without the need for real users.

All in all, there are different ways of measuring an artist's popularity, and might even exist different *domains* of popularity; what is popular in one domain can be unknown in another. As previously stated, popularity is a nebulous concept that can be viewed in different ways.

4.2.3 An Example

Figure 4.2 depicts the Long Tail popularity, using total playcounts, for 260,525 music artists. The horizontal axis contains the list of artists ranked by its total playcounts. For example *The Beatles*, at position 1, has more than 50 million playcounts.

This data was gathered from *last.fm* during July, 2007. *Last.fm* provides plugins for almost any desktop music player (as well as *iPhones* and other mobile

---

[3] http://www.myspace.com/michaelimhof

[4] http://www.myspace.com/curtisyoungofficial

[5] Such as http://www.somanymp3s.com/

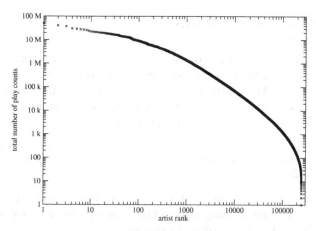

Fig. 4.3 The music Long Tail effect. Same plot as Fig. 4.2 here in log–log scale. The best fit is a log-normal distribution, with a mean of log $\mu = 6.8$, and standard deviation of log, $\sigma = 2.18$. The fast drop in the tail is in part due to misspelled artists (e.g. incorrect metadata in the ID3 tags).

devices) to track users' listening behaviour. It also provides a Flash player embedded in their website, and a client for PC, Mac and Linux that can create personalised audio streams. Figure 4.2 corroborates the music consumption reports by Nielsen Soundscan [5]; a few artists concentrate most of the total plays, whilst many musicians hold the rest. Figure 4.3 presents the same data as Fig. 4.2, in log–log scale. The best fit for the curve is a log-normal distribution, with parameters mean of log $\mu = 6.8$, and standard deviation of log $\sigma = 2.18$ (more information about fitting a curve with a distribution model is presented in Sec. 4.3.2). It is worth noting that the fast drop in the tail is in part due to misspelled artists (e.g. incorrect metadata in the ID3 tags).

4.3 Definitions

The Long Tail of a catalog is measured using the frequency distribution (e.g. purchases, downloads, etc.), ranked by item popularity. We present now two definitions for the Long Tail. The first one is an informal, intuitive one. The second one is a quantitative definition that uses a formal model to characterise the shape of the curve, and a method to fit the data to some well-known distributions (e.g. power-law, power-law with exponential decay, log-normal, etc.).

4.3.1 Qualitative, Informal Definition

According to Chris Anderson [1], the Long Tail is divided in two separate parts: the head and the tail. The head part contains the items one can find in the *old* brick-and-mortar markets. The tail of the curve is characterised by the remainder of the existing products. This includes the items that are available in on-line markets. Chris Anderson's definition, based on the economics of the markets, is:

> *The Long Tail is about the economics of abundance; what happens when the bottlenecks that stand between supply and demand in our culture start to disappear and everything becomes available to everyone.*

The definition emphasises the existence of two distinguished markets; the familiar one (the *Head*), and the long ignored but emerging since the explosion of the web (the *Tail*), consisting of small niche markets.

Another definition is the one by Jason Foster:

> *The Long Tail is the realization that the sum of many small markets is worth as much, if not more, than a few large markets.*[6]

Both definitions are based on markets and economics, and do not propose any computational model to compute and characterise any tail curve, nor fit the data to any existing distribution. Indeed, [1] does not define how to split the head and the tail parts, that are the two key elements in both definitions.

4.3.1.1 Physical Apples Versus Online Oranges

Since *The Long Tail* book became a top-seller, there is a lot of criticism against Anderson's theory. The most common criticism is the lack of scientific backup when comparing different data sources. That is, when comparing the online world to the physical world, Anderson simplifies too much. For instance, he considers only one brick-and-mortar store (e.g. *Walmart*), and compares their music catalog with the one found in the *Rhapsody* online store. However, in the real world there are much more music stores than *Walmart*. Indeed, there are specialised music stores that carry out ten times the volume of *Walmart*'s music catalog. Sadly enough, these ones are completely ignored in Anderson's studies [7].

In addition, there is no clear evidence that online stores can monetise the Long Tail. According to Elberse et al. there is no evidence of a shift in online markets towards promoting the tail [8]. The tail is long, but extremely flat. In their results, hit-driven economies are found in both physical and online markets. Furthermore, in an older study [9], Elberse found that the long tail of movies, those that sell only a few copies every week nearly doubled during their study period. However, the number of non-selling titles rose four times, thus increasing the size of the tail. Regarding the head of the curve; a few mainstream movies still accounted for most of the sales.

---

[6] From http://longtail.typepad.com/the_long_tail/2005/01/definitions _fin.html

Another drawback of the theory is the creation of online oligarchies. "Make everything available" is commonly achieved by *One-Big-Virtual-Tent* rather than *Many-Small-Tents.*[7] That is to say, there is only one *Amazon* that provides most of the content.

Last but not least, Anderson's theory states that the Long Tail follows a power-law distribution. That is a straight line in a log–log plot. However, only plotting a curve in a log–log scale is not enough to verify that the curve follows a power-law. It can better fit to other distributions, such as log-normal or a power-law with an exponential decay of the tail. We need, then, a model that allows us to quantitative define the shape of the Long Tail curve, without the need of linking it with niche markets, economics, and profitable (or not) e-Commerce websites.

4.3.2 Quantitative, Formal Definition

The Long Tail model, $F(x)$, simulates any heavy-tailed distribution [10]. It models the cumulative distribution of the Long Tail data. $F(x)$ represents the share (%) of total volume covered by objects up to rank x:

$$F(x) = \frac{\beta}{(\frac{N_{50}}{x})^\alpha + 1} \tag{4.2}$$

where α is the factor that defines the S-shape of the function, β is the total volume share (and also describes the amount of latent demand), and N_{50}, the median, is the number of objects that cover half of the total volume, that is $F(N_{50}) = 50$.

Once the Long Tail is modelled using $F(x)$, we can divide the curve in three parts: head, mid, and the tail. The boundary between the head and the mid part of the curve is defined by:

$$X_{head \to mid} = N_{50}^{2/3} \tag{4.3}$$

Likewise, the boundary between the mid part and the tail is:

$$X_{mid \to tail} = N_{50}^{4/3} \simeq X_{head \to mid}^2 \tag{4.4}$$

Figure 4.4 depicts the cumulative distribution of the Long Tail of the 260,525 music artists presented in Fig. 4.2. Interestingly enough, the top-737 artists, 0.28% of all the artists, account for 50% of the total playcounts, $F(737) = 50(N_{50} = 737)$, and only the top-30 artists hold around 10% of the plays. Another measure is the *Gini coefficient*. This coefficient measures the inequality of a given distribution, and it determines the degree of imbalance [11]. In our Long Tail example, 14% of the artists hold 86% of total playcounts, yielding a Gini coefficient of 0.72. This value describes a skewed distribution, higher than the classic 80/20 Pareto rule, with a

---

[7] See Tom Slee critical reader's companion to "The Long Tail" book at http://whimsley.typepad.com/whimsley/2007/03/the_long_tail_1.html

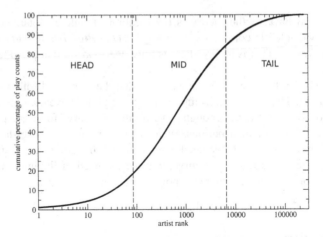

Fig. 4.4 Example of the Long Tail model. It shows the cumulative percentage of playcounts of the 260,525 music artists from Fig. 4.2. Only top-737 artists, 0.28% of all the artists, accumulates the 50% of total playcounts (N_{50}). Also, the curve is divided in three parts: head, mid and tail ($X_{head \rightarrow mid} = 82$, and $X_{mid \rightarrow tail} = 6,655$), so each artist is located in one section of the curve.

value of 0.6. Figure 4.4 also shows the three different sections of the Long Tail. The head of the curve, $X_{head \rightarrow mid}$ consists of only 82 artists, whilst the mid part has 6,573 ($X_{mid \rightarrow tail} = 6,655$). The rest of the artists are located in the tail.

4.3.2.1 Fitting a Heavy-Tailed Distribution Using $F(x)$

To use the $F(x)$ function we need to fit the curve with an estimation of α, β and N_{50} parameters. We do a non-linear regression, using Gauss–Newton method for non-linear least squares, to fit the observations of the cumulative distribution to $F(x)$.[8] Figure 4.5 shows an example of the fitted distribution using the $F(x)$ model. The data is the one from artist popularity in *last.fm* (Fig. 4.4).

4.3.3 Qualitative Versus Quantitative Definition

On the one hand, the qualitative definition by Anderson emphasises the economics of the markets, and the shift from physical to virtual, online, goods. On the other hand, the quantitative definition is based on a computational model that allows us to fit a set of observations (of the cumulative distribution) to a given function, $F(x)$.

The main difference between the two definitions (qualitative and quantitative) is the way each method split the curve into different sections (e.g. the head and

---

[8] To solve the non-linear least squares we use the R statistical package. The code is available at http://mtg.upf.edu/~ocelma/PhD

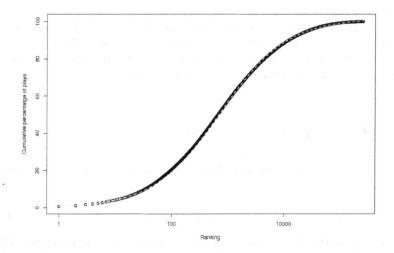

Fig. 4.5 Example of fitting a heavy-tailed distribution (the one in Fig. 4.4) with $F(x)$. The *black dots* represent the observations while the *white dotted curve* represents the fitted model, with parameters $\alpha = 0.73$, and $\beta = 1.02$.

the tail). The qualitative approach is based on the % covered by x (e.g. "20% of the products represent 80% of sales") whereas the quantitative definition splits the x (log) axis equally in three (head, mid, and tail) parts. The main problem is that when adding many more products (say 10,000) in the curve, the changes in the *head* and *tail* boundaries are very radical in the qualitative definition. The quantitative approach does not suffer from this problem. The changes in the section boundaries are not so extreme.

4.4 Characterising a Long Tail Distribution

An early mention of the "long tail", in the context of the Internet, was Clay Shirky's essay in February, 2003.[9] After that, [1] converted the term to a proper noun, and defined a new trend in economics. Since then, the spotlight on the "Long Tail" noun has created many different opinions about it.

In our context, we use a "Long Tail" curve to describe the popularity phenomenon in any recommender system, to show how popularity can affect the recommendations. So, given a long tail distribution of the items' popularity, an important step is to characterise the shape of the curve to understand the amount of skewness.

---

[9] See http://shirky.com/writings/powerlaw_weblog.html

We characterise a Long Tail distribution using Kilkki's $F(x)$ function. Its parameters α, β, and N_{50} defines the shape of the curve.

Yet, it is also important to determine the shape of the curve according to well-known probability density distributions. There are different probability density distribution functions that can fit a heavy-tailed curve. We present some of them here: power-law, power-law with exponential decay, and log-normal distribution.

4.4.1 Not All Long Tails Are Power-Law

A *power-law distribution* is described using the probability density distribution (*pdf*), $f(x)$:

$$f(x) = ax^{-\gamma} \tag{4.5}$$

Power-law distribution has the property of (asymptotic) scale invariance. This type of distribution cannot be entirely characterised by its mean and variance. Also, if the γ power-law exponent has a value close to 1, $\gamma \simeq 1$, then this means that the long tail is fat.[10] In other words, a power-law with $\gamma \gg 1$ consists of a thin tail (with values close to 0), and a short head with a high probability value.

Power-law with an exponential decay distribution differs from a power-law by the shape of the tail. Its *pdf* is defined by:

$$f(x) = x^{-\gamma}e^{-\lambda x}, \tag{4.6}$$

There exists an N that denotes the threshold between the power-law distribution $(x^{-\gamma}, x \leq N)$, and the exponential decay $(e^{-\lambda x}, x > N)$. This means that the tail of the curve is better represented with an exponential cut-off.

In a *log-normal* distribution the logarithm of the variable is normally distributed. That is to say, if a variable X is normally distributed, then $Y = e^X$ has a log-normal distribution. Log-normal distribution promotes the head of the curve. It is a distribution skewed to the right, where the popular items have a strong effect, whilst the tail has a very small contribution in the *pdf*:

$$f(x) = \frac{1}{x}e^{-\frac{(\ln(x)-\mu)^2}{2\sigma^2}} \tag{4.7}$$

Thus, the main problem is, given a curve in a log–log scale representation, to decide which is the best model that explains the curve. It is worth noting that, according to Anderson's theory (i.e. the Long Tail is profitable), the curve should be modelled as a power-law, with $\gamma \simeq 1$, meaning that the tail is fat. However, if the best fit is using another distribution, such as a log-normal—which is very common— then Anderson's theory cannot be strictly applied in that particular domain, and context.

---

[10] This is the only case where Anderson's Long Tail theory can be applied.

4.4.2 A Model Selection: Power-Law or Not Power-Law?

To characterise a heavy-tailed distribution, we follow the steps described in Clauset et al. [12]. As previously mentioned, the main drawbacks when fitting a Long Tail distribution are: (i) to plot the distribution on a log–log plot, and see whether it follows a straight line or not, and (ii) use linear regression by least squares to fit a line in the log–log plot, and then use R^2 to measure the fraction of variance accounted for the curve. This approach gives a poor estimate of the model parameters, as it is meant to be applied to regression curves, not to compare distributions. Instead, to decide whether a heavy-tailed curve follows a power-law distribution, [12] propose the following steps:

1. *Estimate γ.* Use the maximum likelihood estimator (MLE) for the γ scaling exponent. MLE always converge to the correct value of the scaling exponent.
2. *Detect x_{min}.* Use the goodness of fit value to estimate where the scaling region begins (x_{min}). The curve can follow a power-law on the right or upper tail, so above a given threshold x_{min}. The authors propose a method that can empirically find the best scaling region, based on the Kolmogorov–Smirnov D statistic.
3. *Goodness of the model.* Use, again, the Kolmogorov–Smirnov D statistic to compute the discrepancy between the empirical distribution and the theoretical one. The Kolmogorov–Smirnov (K–S) D statistic will converge to zero, if the empirical distribution follows the theoretical one (e.g. power-law). The K–S D statistic for a given cumulative distribution function $F(x)$, and its empirical distribution function $F_n(x)$ is:

$$D_n = \sup_x |F_n(x) - F(x)|, \qquad (4.8)$$

where $\sup |S|$ is the supremum of a set S. That is the lowest element of S that is greater than or equal to each element of S. The supremum is also referred to as the *least upper bound*.
4. *Model selection.* Once the data is fitted to a power-law distribution, the only remaining task is to check among the different alternatives. That is, to detect whether other non power-law distributions could have produced the data. This is done using pairwise comparison (e.g. power-law versus power-law with exponential decay, power-law versus a log-normal, etc.), and [12] use the Vuong's test [13]. Vuong's test uses the log-likelihood ratio and the Kullback–Leibler information criterion to make probabilistic statements about the two models. Vuong's statistical test is used for the model selection problem, where one can determine which distribution is closer to the real data. A large, positive Vuong's test statistic provides evidence of the best fitting using a power-law distribution over the other distribution, while a large, negative test statistic is an evidence of the contrary.

4.5 The Dynamics of the Long Tail

Another important aspect of any Long Tail is its dynamics. E.g., does an artist stay
in the head region forever? Or the other way around; will niche artists always remain
in the long tail? Figure 4.6 depicts the increase of the Long Tail popularity after 6
months, using 50,000 out of the 260,525 *last.fm* artists (see Fig. 4.2). Figure 4.6
shows the dynamics of the curve comparing two snapshots; one from July 2007,
and the other from January 2008. The most important aspect is the increase of total
playcounts in each area of the curve.

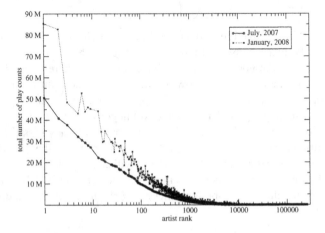

Fig. 4.6 The dynamics of the Long Tail after 6 months (between July, 2007 and January, 2008).
Radiohead, at top-2, is now closer to *The Beatles* (top-1), due to the release of their *In Rainbows*
album.

4.5.1 Strike a Chord?

Table 4.4 shows the playcount increment, in %. In all the three regions—head, mid,
and tail—the percentage increment of plays is almost the same (around 62%), meaning that not many artists move between the regions. For instance, in the head area,
Radiohead at top-2 is much closer to top-1, *The Beatles*, due to the release of the
In Rainbows album.[11] Still, the band remains at top-2. An interesting example in
the tail area is the *Nulla Costa* band. This band was at rank 259,962 in July, 2007.
After 6 months they increase from 3 *last.fm* playcounts to 4,834, positioning at rank
55,000. Yet, the band is still in the tail region. We could not detect any single artist

---

[11] *In Rainbows* album was released on October 10th, 2007

that clearly moved from the tail to the mid region.[12] There exist niche artists, and the main problem is to find them. The only way to leverage the long tail is by providing recommendations that promote unknown artists.

Once the Long Tail is formally described, the next step is to use this knowledge when providing recommendations. The following section presents how one can exploit the Long Tail to provide novel or familiar recommendations, taking into account the user profile.

Long Tail region	Increase (%)
Head	61.20
Mid	62.29
Tail	62.32

Table 4.4 Increase of the Long Tail regions (in %) after 6 months (comparing two snapshots in July, 2007 and January, 2008).

4.6 Novelty, Familiarity and Relevance

If you like The Beatles you might like...X. Now, ask several different people and you will get lots of different $X's$. Each person, according to her ties with the band's music, would be able to propose interesting, surprising or expected $X's$. Nonetheless, asking the same question to different recommender systems we are likely to get similar results. Indeed, two out of five tested music recommenders contain John Lennon, Paul McCartney and George Harrison in their top-10 (*last.fm* and *the.echotron.com* by *The Echo Nest* company). *Yahoo! Music* recommends John Lennon and Paul McCartney (1st and 4th position), whereas *Mystrands.com* only contains John Lennon (at top-10). Neither *ilike* nor *Allmusic.com* contain any of these musicians in their list of *Beatles'* similar artists. Furthermore, Amazon's top-30 recommendations for the *Beatles' White Album* is strictly made of other *Beatles'* albums (all of a sudden, at the fourth page of the navigation there is the first non-*Beatles* album; *Exile on Main St.* by The Rolling Stones). Finally, creating a playlist from *OneLlama.com*— starting with a *Beatles* seed song—one gets four out of ten songs from the *Beatles*, plus one song from John Lennon, so it makes half of the playlist. It is worth mentioning that these recommenders use different approaches, such as: collaborative filtering, social tagging, web mining and co-occurrence analysis of playlists. To conclude this informal analysis, the most noticeable fact is that only *last.fm* remembers Ringo Starr![13]

---

[12] *Last.fm* has the "hype artist" weekly chart, http://www.last.fm/charts/hypeartist, a good source to track the movements in the Long Tail curve.

[13] This informal analysis was done in July, 2007.

One can agree or disagree with all these *Beatles'* similar artist lists. However, there are a very few, if none at all, serendipitous recommendations (the rest of the similar artists were, in no particular order: *The Who, The Rolling Stones, The Beach Boys, The Animals,* and so on). Indeed, some of the before mentioned systems provide filters, such as: "surprise me!" or the "popularity slider", to dive into the Long Tail of the catalog. Thus, novel recommendations are sometimes necessary to improve the user's experience and discovery in the recommendation workflow.

It is not our goal to decide whether one can monetise the Long Tail or to exploit the niche markets, but to help people discover those items that are lost in the tail. Hits exist and they always will. Our goal is to motivate and guide the discovery process, presenting to users rare, non-hit, items they could find interesting.

4.6.1 Recommending the Unknown

It has been largely acknowledged that item popularity can decrease user satisfaction by providing obvious recommendations [14, 15]. Yet, there is no clear recipe for providing *good* and *useful* recommendations to users. We can foresee at least three key aspects that should be taken into account. These are: novelty and serendipity, familiarity, and relevance [16]. According to Wordnet dictionary,[14] **novel** (*adj.*) has two senses: "new—original and of a kind not seen before"; and "refreshing—pleasantly new or different". Serendipity (*noun*) is defined as "good luck in making unexpected and fortunate discoveries". Familiar (*adj.*) is defined as "well known or easily recognised". In our context, we measure the novelty for a given user u as the ratio of unknown items in the list of top-N recommended items, $\mathcal{L}_N$:

$$Novelty(u) = \frac{\sum_{i \in \mathcal{L}_N}(1 - Knows(u,i))}{N}, \qquad (4.9)$$

being $Knows(u,i)$ a binary function that returns 1 if user u already knows item i, and 0 otherwise. Likewise, user's familiarity with the list of recommended items can be defined as $Familiar(u) = 1 - Novelty(u)$.

Nonetheless, a user should be familiar with some of the recommended items, to improve confidence and trust in the system. Also, some items should be unknown to the user (discovering *hidden* items in the catalog). A system should also give an explanation of why those—unknown—items were recommended, providing a higher confidence and transparency on these recommendations. The difficult job for a recommender is, then, to find the proper level of familiarity, novelty and relevance for *each* user.

Figure 4.7 shows the long tail of item popularity, and it includes a user profile. The profile is exhibited as the number of times the user has interacted with that item. Taking into account item popularity plus the user profile information, a

---

[14] http://wordnet.princeton.edu

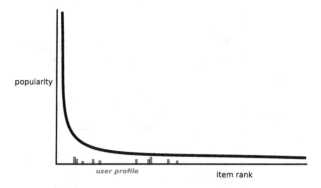

Fig. 4.7 A user profile represented in the Long Tail. The profile is exhibited as the number of times the user has interacted with that item.

recommender can provide personalised and relevant recommendations that are also novel to the user.

4.6.1.1 Trade-Off Between Novelty and Relevance

However, there is a trade-off between novelty and user's relevance. The more novel, unknown items a recommender presents to a user, the less relevant they can be perceived by her.

Figure 4.8 presents the trade-off between novelty and relevance. It shows the different recommendation states for a given a user u, given a large collection of items (say, not only the user's personal collection). The gray triangle represents the area where a recommender should focus on to provide relevant items to u. On the one hand, laid-back recommendations (bottom-right) appear when the system recommends familiar and relevant items to u. On the other hand, the discovery process (top-right) starts when the system provides to the user (potentially) unknown items that could fit in her profile. The provided recommendations should conform to the user's intentions; sometimes a user is expecting familiar recommendations (laid-back state), while in other cases she is seeking to actively discovery new items.

There are two more cases, that is when the recommender provides popular items, and when it provides random ones. This can happen when there is not enough information about the user (e.g. the user cold-start problem). In this case, the system can recommend popular items (bottom-left). Popular items are expected to be somehow familiar to the user, but not necessarily relevant to her. The other situation is when the system provides random recommendations to u (top-left). This case is similar to a shuffle playlist generator, with the difference that in our case the items' catalog is much bigger than the personal music collection of u. Thus, there is less chances that user u might like any of the random recommendations, as they are not personalised at all.

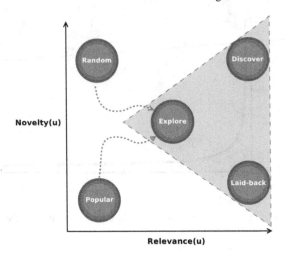

Fig. 4.8 Trade-off between novelty and relevance for a user *u*.

4.6.2 Related Work

Serendipity and novelty are relevant aspects in the recommendation workflow [15]. Indeed, there is some related work that explicitly addresses these aspects. For instance, five measures to capture redundancy are presented in [17]. These measures allow one to infer whether an item—that is considered relevant—contains any novel information to the user. Yang and Li [18] defines novelty in terms of the user knowledge and her degree of interest in a given item. In [19], Weng et al. propose a way to improve the quality and novelty of the recommendations by means of a topic taxonomy-based recommender, and hot topic detection using association rules. Other proposals include disregarding items if they are too similar to other items that the user has already seen [20], or simple metrics to measure novelty and serendipity based on the average popularity of the recommended items [21].

Even though all these approaches focus on providing novel and serendipitous recommendations, there is no framework that consistently evaluates the provided recommendations. Thus, there is a need to design evaluation metrics to deal with the effectiveness of novel recommendations, not only measuring prediction accuracy, but taking into account other aspects such as usefulness and quality [14, 22]. Novelty metrics should look at how well a recommender system made a user aware of previously unknown items, as well as to what extent users accept the new recommendations [14].

Generally speaking, the most popular items in the collection are the ones with higher probability that a given user will recognise, or be broadly familiar with. Likewise, one can assume that items with less interaction—rating, purchasing, previewing—within the community of users are more likely to be unknown [21]. In this sense, the Long Tail of the items' catalog assists us in deciding how novel or familiar an item could be. Yet, a recommender system must predict whether an item could be relevant, and then be recommended, to a user.

4.7 Summary

Effective recommendation systems should promote novel and relevant material (non-obvious recommendations), taken primarily from the tail of a popularity distribution. In this sense, the Long Tail can be described in terms of niche markets' economics, but also by describing the item popularity curve. We use the latter definition—the Long Tail model, $F(x)$—to describe the cumulative distribution of the curve. In the music field, the $F(x)$ model allows us to define artist popularity, and her location in the curve (head, mid or tail region). Hence, $F(x)$ denotes the shared knowledge about an artist, by a community of listeners. From this common knowledge, we can derive whether an artist can be novel and relevant to a given user profile.

Our results show that music listening habits follow the hit-driven (or mainstream) paradigm; 0.28% (737 out of 260,525) of the artists account for the 50% of total playcounts. The best fit (in the log–log plot) for the music Long Tail is a log-normal distribution. A log-normal distribution concentrates most of the information in the the head region. Even though we use playcounts and not total sales to populate the curve, this finding unveils Anderson's theory about the economics and monetisation in the Long Tail. Despite Anderson's failure or success theory, his core idea still is an interesting way to explain the changes the web has provoked, in terms of the availability of all kind of products—from hits to niches.

One of the goals of a recommender should be to promote the tail of the curve by providing relevant, personalised novel recommendations to its users. That is, to smoothly interconnect the head and mid regions with the tail, so the recommendations can drive interest from one to the other. Figure 4.9 presents this idea. It depicts a 3D representation of the Long Tail; showing the item popularity curve, the similarities among the items, and a user profile denoted by her preferred items (in dark gray colour). The set of candidate items (dotted lines) to be recommended to the user are shown also. Items' height denotes the relevance for that user. Candidate items located in the tail part are considered more novel—and, potentially relevant—than the ones in the head region.

4.7.1 Links with the Following Chapters

In this chapter we have presented the basics for novelty detection in a recommender system, using the popularity information and its Long Tail shape. The next step is to evaluate these types of recommendations. We can foresee two different ways to evaluate novel recommendations, and these are related with (i) exploring the available (and usually, very large) item catalog, and (ii) filtering new incoming items.

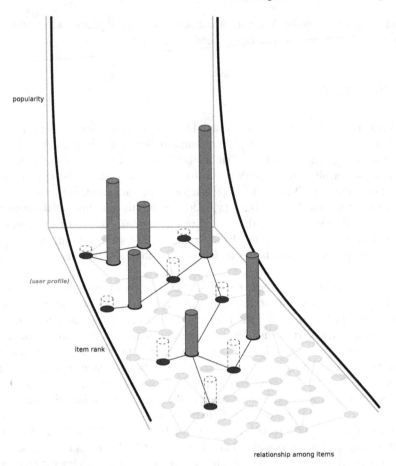

Fig. 4.9 A 3D representation of the Long Tail. It adds another dimension; the similarities among the items, including the representation of a user profile (*in gray*). The set of candidate items to be recommended to the user are shown (*in dotted lines*) and its height denotes the relevance for the user.

We mainly focus on the former case, and we present two complementary evaluation methods. On the one hand, *network-centric* evaluation method (presented in Chap. 6) focuses on analysing the items' similarity graph, created using any item-based recommendation algorithm. The aim is to detect whether the intrinsic topology of the items' network has any pathology that hinders novel recommendations, promoting the most popular items. On the other hand, a *user-centric* evaluation aims at measuring the perceived quality of novel recommendations. This user evaluation is presented in Chap. 7. Yet, before presenting the evaluation results we introduce, in Chap. 5, the metrics that we use.

References

1. C. Anderson, *The Long Tail. Why the Future of Business Is Selling Less of More*. New York, NY: Hyperion, 2006.
2. D. M. Fleder and K. Hosanagar, "Blockbuster culture's next rise or fall: The impact of recommender systems on sales diversity," *SSRN eLibrary*, 2007.
3. C. Tucker and J. Zhang, "How does popularity information affect choices? theory and a field experiment," *SSRN eLibrary*, 2008.
4. M. J. Salganik, P. S. Dodds, and D. J. Watts, "Experimental study of inequality and unpredictability in an artificial cultural market," *Science*, vol. 311, pp. 854–856, February 2006.
5. N. Soundscan, "State of the industry," *Nielsen Soundscan Report. National Association of Recording Merchandisers*, 2007.
6. N. Soundscan, "Nielsen soundscan report. Year–end music industry report," White Plains, NY, 2006.
7. T. Slee, "A critical reader's companion to the long tail," 2006.
8. A. Elberse, "Should you invest in the long tail?" *Harvard Business Review*, vol. 86, no. 7/8, pp. 88–96, 2008.
9. A. Elberse and F. Oberholzer-Gee, "Superstars and underdogs: An examination of the long tail phenomenon in video sales," *Harvard Business School Working Paper*, May 2006.
10. K. Kilkki, "A practical model for analyzing long tails," *First Monday*, vol. 12, May 2007.
11. C. Gini, "Measurement of inequality and incomes," *The Economic Journal*, vol. 31, pp. 124–126, 1921.
12. A. Clauset, C. R. Shalizi, and M. E. J. Newman, "Power-law distributions in empirical data," *SIAM Reviews*, June 2007.
13. Q. H. Vuong, "Likelihood ratio tests for model selection and non-nested hypotheses," *Econometrica*, vol. 57, pp. 307–333, March 1989.
14. J. L. Herlocker, J. A. Konstan, L. G. Terveen, and J. T. Riedl, "Evaluating collaborative filtering recommender systems," *ACM Transaction on Information System*, vol. 22, no. 1, pp. 5–53, 2004.
15. S. M. McNee, J. Riedl, and J. A. Konstan, "Being accurate is not enough: How accuracy metrics have hurt recommender systems," in *Computer Human Interaction. Human factors in computing systems*, (New York, NY), pp. 1097–1101, ACM, 2006.
16. O. Celma and P. Lamere, "Music recommendation tutorial," in *Proceedings of 8th International Conference on Music Information Retrieval*, (Vienna, Austria), 2007.
17. Y. Zhang, J. Callan, and T. Minka, "Novelty and redundancy detection in adaptive filtering," in *Proceedings of the 25th International ACM SIGIR Conference on Research and Development in Information Retrieval*, (New York, NY), pp. 81–88, ACM, 2002.
18. Y. Yang and J. Z. Li, "Interest-based recommendation in digital library," *Journal of Computer Science*, vol. 1, no. 1, pp. 40–46, 2005.
19. L.-T. Weng, Y. Xu, Y. Li, and R. Nayak, "Improving recommendation novelty based on topic taxonomy," in *Proceedings of the IEEE/WIC/ACM International Conferences on Web Intelligence and Intelligent Agent Technology*, (Washington, DC), pp. 115–118, IEEE Computer Society, 2007.
20. D. Billsus and M. J. Pazzani, "User modeling for adaptive news access," *User Modeling and User-Adapted Interaction*, vol. 10, no. 2–3, pp. 147–180, 2000.
21. C.-N. Ziegler, S. M. McNee, J. A. Konstan, and G. Lausen, "Improving recommendation lists through topic diversification," in *Proceedings of the 14th International Conference on World Wide Web*, (New York, NY), pp. 22–32, ACM, 2005.
22. G. Adomavicius and A. Tuzhilin, "Toward the next generation of recommender systems: A survey of the state-of-the-art and possible extensions," *IEEE Transactions on Knowledge and Data Engineering*, vol. 17, no. 6, pp. 734–749, 2005.

Chapter 5
Evaluation Metrics

This chapter presents the different evaluation methods for a recommender system. We introduce the existing metrics, as well as the pros and cons of each method. This chapter is the background for the following Chap. 6 and 7, where the proposed metrics are used in real, large size, recommendation datasets.

5.1 Evaluation Strategies

We classify the evaluation of recommender algorithms in three groups; system-, network-, and user-centric.

- *System-centric* evaluation measures how accurate the system can predict the actual values that user have previously assigned. This approach has been extensively used in collaborative filtering with explicit feedback (e.g. ratings).
- *Network-centric* evaluation aims at measuring the topology of the item (or user) similarity network. It uses metrics from complex network analysis (CNA). Network-centric evaluation measures the inherent structure of the item (or user) similarity network. The similarity network is the basis to provide the recommendations. Thus, it is important to analyse and understand the underlying topology of the similarity network.
- *User-centric* evaluation focuses on the user's perceived quality and usefulness of the recommendations. This evaluation requires the user intervention —via survey, or gaterhing information from the user activity in the system.

The following sections are devoted to explain each evaluation method.

Ò. Celma, *Music Recommendation and Discovery*,
DOI 10.1007/978-3-642-13287-2_5, © Springer-Verlag Berlin Heidelberg 2010

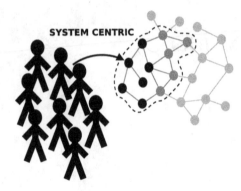

Fig. 5.1 System-centric evaluation is based on the analysis of the subcollection of items of a user, using the *leave-n-out* method [1], and aggregating (e.g. averaging) the results for all users to provide a final compact metric.

5.2 System-Centric Evaluation

As of today, system-centric evaluation has been largely applied. The most common approaches are based on the *leave-n-out* method [1], that resembles to the classic *n-fold* cross validation. Given a dataset where a user has implicitly or explicitly interacted with (via ratings, purchases, downloads, previews, etc.), split the dataset in two—usually disjunct—sets: training and test. Accuracy evaluation is based only on a user's dataset, so the rest of the items of the catalog are ignored. Figure 5.1 presents the method.

The evaluation process includes, then, several metrics such as: predictive accuracy (Mean Absolute Error, Root Mean Square Error), decision based (Mean Average Precision, Recall, F-measure, and ROC), and rank based metrics (Spearman's ρ, Kendall-τ, and half-life utility) [2, 3].

5.2.1 Predictive-Based Metrics

Predictive metrics aim at comparing the predicted values against the actual values. The result is the average over the deviations.

5.2.1.1 Mean Absolute Error (MAE)

Given a test set $\mathcal{T}$ of user-item pairs (u, i) with ratings $r_{u,i}$, the system generates predicted ratings $\hat{r}_{u,i}$. Mean Absolute Error (MAE) measures the deviation between the predicted value and the real value:

$$MAE = \frac{1}{|\mathcal{T}|} \sum_{(u,i)\in\mathcal{T}} |\hat{r}_{u,i} - r_{u,i}|, \tag{5.1}$$

where $\hat{r}_{u,i}$ is the predicted value of user u for item i, and $r_{u,i}$ the true value.

5.2.1.2 Root Mean Squared Error (RMSE)

Mean Squared Error (MSE) is also used to compare the predicted value with the real value a user has assigned to an item. The difference between MAE and MSE is that MSE heavily penalise large errors.

$$MSE = \frac{1}{|\mathcal{T}|} \sum_{(u,i)\in\mathcal{T}} (\hat{r}_{u,i} - r_{u,i})^2 \tag{5.2}$$

Root Mean Squared Error (RMSE) equals to the square root of the MSE value.

$$RMSE = \sqrt{MSE} \tag{5.3}$$

RMSE is one of the most used metrics in collaborative filtering based on explicit ratings. RMSE is the metric that was used in the Netflix \$1,000,000 prize.

Related metrics are Average RMSE and Average MAE. In this case, we compute the RMSE (or MAE) for each item and then take the average over all items. Likewise, we can compute the RMSE (or MAE) separately for each user and then take the average over all users.

5.2.2 Decision-Based Metrics

Decision-based metrics evaluates the top-N recommendations for a user. Recommendations comes in a ranked list of items, ordered by decreasing relevance. There are four different cases to take into account:

- True positive (TP). The system recommends an item the user is interested in.
- False positive (FP). The system recommends an item the user is not interested in.
- True negative (TN). The system does not recommend an item the user is not interested in.
- False negative (FN). The system does not recommend an item the user is interested in.

Precision (P) and recall (R) are obtained from the 2×2 contingency table (or confusion matrix) shown in Table 5.1. The recommended items are separated into two classes; relevant or not relevant according to the user profile. When the rating scale is not binary, we need to transform it into a binary scale, to decide whether the item is relevant or not. E.g. in a rating scale of [1..5], ratings of 4 or 5 are considered relevant, and ratings from 1..3 as not-relevant.

	Relevant	Not relevant
Recommended	TP	FP
Not recommended	FN	TN

Table 5.1 Contingency table showing the categorisation of the recommended items in terms of relevant or not. Precision and recall metrics are derived from the table.

5.2.2.1 Precision

Precision measures the fraction of relevant items over the recommended ones.

$$Precision = \frac{TP}{TP+FP} \tag{5.4}$$

Precision can also be evaluated at a given cut-off rank, considering only the top-n recommendations. This measure is called precision-at-n or P@n.

5.2.2.2 Recall

The recall measures the coverage of the recommended items, and is defined as:

$$Recall = \frac{TP}{TP+FN} \tag{5.5}$$

Recall is also known as sensitivity, true positive rate (TPR), or hit-rate.

5.2.2.3 F-Measure

F-measure combines P and R results, using the weighted harmonic mean. The general formula (for a non-negative real β) is:

$$F_\beta = \frac{(1+\beta^2)\cdot(\text{precision}\cdot\text{recall})}{(\beta^2\cdot\text{precision}+\text{recall})} \tag{5.6}$$

Two common F-measures are F_1 and F_2. In F_1 recall and precision are evenly weighted, and F_2 weights recall twice as much as precision.

5.2.2.4 Accuracy

Accuracy is the simplest way to evaluate the predicted recommendations. Accuracy measures the ratio of correct predictions versus the total number of items evaluated. Accuracy is also obtained from the 2×2 contingency table.

$$Accuracy = \frac{TP+TN}{TP+FP+TN+FN} \tag{5.7}$$

5.2.2.5 Receiver Operating Characteristic (ROC) Curve

The previous decision-based metrics (P, R, F-measure) use a fixed recommendation list length. Receiver Operating Characteristic (ROC) curve measures the selection of high-quality items over a range of different recommendation list lengths, for a given user. ROC measures the trade-off between hit-rates (TPR) and false-alarm rates (or false positive rates, FPR). Hit-rate, or True Positive Rate, is defined as:

$$TPR = Recall = \frac{TP}{TP+FN} \tag{5.8}$$

False positive rate (FPR) equals to:

$$FPR = \frac{FP}{FP+TN} \tag{5.9}$$

ROC can visualise the trade-off between TPR and FPR. The random curve assigns a probability of 50% to each of the two classes (recommended, not recommended). The area under the curve (AUC) is a measure that summarises a ROC result. A random curve has an AUC of 0.5. The closer the AUC to 1, the better.

The main drawback of all the previous decision-based metrics is that do not take into account the ranking of the recommended items. Thus, item at top-1 has the same relevance as item at top-20. To avoid this limitation, one can use rank-based metrics.

5.2.3 Rank-Based Metrics

There are two approaches to evaluate ranked lists of recommendations. The first one it to determine the order of the predicted items for a given user, and compare this with the correct order, or reference ranking. The second approach is to measure the utility of the predicted list. In this case, items in the top positions are considered more rellevant than the ones in the bottom of the list (e.g. the bottom of a webpage). Whenever the list of recommendations is very large, a pagination is also provided, so the user can browse the whole list of recommended items.

When using a *reference ranking* to compare against, the following measures can be applied: Spearman's rho (ρ), Kendall–tau (τ), and Normalised distance-based performance (NDPM).

5.2.3.1 Spearman's Rho (ρ)

Spearman's ρ computes the rank-based Pearson correlation of two ranked lists. It compares the predicted list with the user preferences (e.g. the ground truth data). Spearman's ρ is defined as:

$$\rho = \frac{1}{n_u} \frac{\sum_i (r_{u,i} - \bar{r})(\hat{r}_{u,i} - \hat{\bar{r}})}{\sigma(r)\sigma(\hat{r})} \tag{5.10}$$

where $\bar{r}$ and $\sigma(\cdot)$ denote the mean and standard deviation, and n_u the number of items for user u.

5.2.3.2 Kendall-Tau (τ)

Kendall-τ also compares the recommended list with the user's preferred list of items. Kendall-τ rank correlation coefficient is defined as:

$$\tau = \frac{C^+ - C^-}{\frac{1}{2}n(n-1)} \tag{5.11}$$

where C^+ is the number of concordant pairs, and C^- is the number of discordant pairs in the data set.

5.2.3.3 Normalised Distance-Based Performance

Normalised Distance-based Performance metric (NDPM) was introduced in [4] to evaluate their collaborative filtering recommender system, named *FAB*.

NDPM is a normalised distance ($[0..1]$), between the user's classification for a set of documents and the system's classification for the same documents [5]. In recommender systems, NDPM measures the difference between a user's and the system's choices. NDPM is defined as:

$$NDPM = \frac{2C^- + C^u}{2C^i} \tag{5.12}$$

where C^- is number of mismatched preference relations between the system and user rankings, C^u is the number of compatible preference relations, and C^i is the total number of preferred relationships in the user's ranking.

The previous metrics compares two ranked lists, but do not take into account its utility. Top items are considered more relevant than items in the bottom of the recommendation list. *Utility-based ranking* metrics take into account the item position in the predicted list of recommendations.

5.2.3.4 Average Reciprocal Hit-Rate

Average Reciprocal Hit-Rate (ARHR) is defined as:

$$ARHR = \frac{1}{n} \sum_{i=1}^{h} \frac{1}{p_i} \tag{5.13}$$

where h is the number of hits that occurred at positions $p_1, p_2, ..., p_h$ within the top-n list. Hits that occur earlier in the top-n list are weighted higher than hits that occur later in the list. ARHR rewards each hit based on where is located in the top-N list. It resembles to the Mean Reciprocal Rank metric from Information Retrieval.

5.2.3.5 Half-Life Utility

Half-life utility (or R-Score) metric attempts to evaluate the utility of the predicted list of items [1]. The utility is defined as the deviation between a user's rating and the default rating for an item. So, half-life utility can be used in algorithms that are based on user explicit feedback, such as ratings. Breese et al. describe the likelihood that a user will view each successive item in the ranked list with an exponential decay function. The strength of the decay is described by a half-life parameter α [1]. Half-life utility is defined as:

$$R_u = \sum_i \sum_j \frac{max(r_{i_j} - d, 0)}{2^{\frac{j-1}{\alpha-1}}} \tag{5.14}$$

where, r_{i_j} represents the rating of user u on item i_j (in the j-position of the ranked list), d is the default rating, and α is the half-life parameter.

5.2.3.6 Discounted Cumulative Gain

Discounted cumulative gain penalises relevant predicted items that are located in the bottom of the recommendation list (e.g. these items should be on top).

$$DCG_p = rel_1 + \sum_{i=2}^{p} \frac{rel_i}{\log_2 i} \tag{5.15}$$

where rel_i is the graded relevance of the recommended item at position i.

5.2.4 Limitations

The main limitation of system-centric evaluation is the set of items that can evaluate. System-centric evaluation cannot avoid the selection bias of the dataset. Users do not rate all the items they receive, but rather they select the ones to rate. The observations a system-centric approach can evaluate is a skewed, narrowed and unrepresentative population of the whole collection of items. That is, for a given user, the system-centric approach only evaluates the items the user has interacted with, neglecting the rest of the collection. The same procedure is applied for the rest of the users, and the final metrics are averaged over all the users.

System-centric metrics present some drawbacks:

- The **coverage** of the recommended items cannot be measured. The collection of items used in the evaluation is limited to the set of items that a user has interacted with.
- The **novelty** of the recommendations cannot be measured. System-centric evaluates the set of items a user has interacted with. Thus, it cannot evaluate the items that are outside this set. Some of these items could be unknown, yet relevant, to the user.
- Neither **transparency** (explainability) nor **trustworthiness** (confidence) of the recommendations can be measured using system-centric metrics.
- The **perceived quality** of the recommendations cannot be measured. Usefulness and effectiveness of the recommendations are two very important aspects for the users. However, system-based metrics cannot measure user satisfaction.

Other user-related elements aspects that a system-centric approach cannot evaluate are the *eclecticness* (preference for disparate and dissimilar items), and *mainstreamness* (preference for popular items) of a user.

To summarise, system-centric metrics evaluate how well a recommender system can predict items that are already in a user profile (assuming that the profile is splited during the train and test steps). The most difficult part is to develop evaluation metrics to deal with the effectiveness of the recommendations. That is, not only measuring prediction accuracy, but taking into account other aspects such as usefulness and quality [6]. Indeed, accuracy is not correlated with the usefulness and subjective quality of the recommendations [7].

5.3 Network-Centric Evaluation

Network-centric evaluation measures the inherent structure of the item (or user) similarity network. The similarity network is the basis to provide the recommendations. Thus, it is important to analyse and understand the underlying topology of the similarity network.

Network-centric evaluation complements the metrics proposed in the system-centric approach. It actually measures other components of the recommender system, such as the coverage, or diversity of the recommendations. However, it only focuses on the collection of items, so the user stays outside the evaluation process. Figure 5.2 depicts this idea.

5.3.1 Complex Network Analysis

We propose several metrics to analyse a recommendation graph; $G := (V, E)$, being V a set of nodes, and E a set of unordered pairs of nodes, named edges. The items (or users) are nodes, and the edges denote the (weighted) similarity among them, using any recommendation algorithm. When using the item similarity graph, we focus on the algorithms that use item-based neighbour similarity. On the other hand,

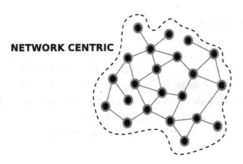

Fig. 5.2 Network-centric evaluation determines the underlying topology of the item (or user) similarity network.

the user similarity graph is the basis for the algorithms that use user-based neighbour similarity. It is worth mentioning that in either case, the similarity network can be created using any recommendation method (e.g. collaborative filtering, content-based, hybrid, etc.). All the proposed metrics are derived from Complex Network and Social Network analysis.

5.3.2 Navigation

5.3.2.1 Average Shortest Path

The average shortest path (or mean geodesic length) measures the distance between two vertices i and j. They are connected if one can go from i to j following the edges in the graph. The path from i to j may not be unique. The minimum path distance (or geodesic path) is the shortest path distance from i to j, d_{ij}. The average shortest path in the network is:

$$\langle d \rangle = \frac{1}{\frac{1}{2}n(n+1)} \sum_{i,j \in V, i \neq j} d_{ij} \qquad (5.16)$$

In a random graph, the average path approximates to:

$$\langle d_r \rangle \sim \frac{logN}{log\langle k \rangle}, \qquad (5.17)$$

where $N = |V|$, and $\langle k \rangle$ denotes the mean degree of all the nodes.

The longest path in the network is called its *diameter* (D). In a recommender system, average shortest path and diameter inform us about the global navigation through the network of items.

5.3.2.2 Giant Component

The strong giant component, *SGC*, of a network is the set of vertices that are connected via one or more geodesics, and are disconnected from all other vertices.
Typically, networks have one large component that contains most of the vertices. It
is measured as the % of nodes that includes the giant component. In a recommender
system, *SGC* informs us about the catalog coverage, that is the total percentage of
available items the recommender recommends to users [2].

5.3.3 Connectivity

5.3.3.1 Degree Distribution

The degree distribution, p_k, is the number of vertices with degree k:

$$p_k = \sum_{v \in V \,|\, \deg(v)=k} 1, \tag{5.18}$$

where v is a vertex, and $\deg(v)$ its degree. More frequently, the *cumulative degree
distribution* (the fraction of vertices having degree k or larger), is plotted:

$$P_c(k) = \sum_{k'=k}^{\infty} p_{k'} \tag{5.19}$$

A cumulative plot avoids fluctuations at the tail of the distribution and facilitates
the computation of the power coefficient γ, if the network follows a power law.
$P_c(k)$ is, then, usually plotted as the complementary cumulative distribution function
(*ccdf*). The complementary cumulative distribution function, $F_c(x)$, is defined as:

$$F_c(x) = P[X > x] = 1 - F(x) \tag{5.20}$$

where $F(x)$ is the cumulative distribution function (*cdf*):

$$F(x) = P[X \leq x] \tag{5.21}$$

$F(x)$ can be regarded as the proportion of the population whose value is less
than x. Thus, $P_c(k)$, derived from $F_c(x)$, denotes the fraction of nodes with a degree
greater than or equal to k.

In a directed graph, that is when a recommender algorithm only computes the
top-n most similar items, $P(k_{in})$ and $P(k_{out})$, the cumulative incoming (outcoming)
degree distribution, are more informative. Complementary cumulative indegree distribution, $P_c(k_{in})$, detects whether a recommendation network has some nodes that
act as hubs. That is, that they have a large amount of attached links. This clearly
affects the recommendations and navigability of the network.

Also, the shape of the curve helps us to identify the network's topology. Regular networks have a constant distribution, "random networks" have a Poisson degree distribution [8] meaning that there are no hubs, and "scale-free networks" follow a power-law distribution in the cumulative degree distribution [9], so there are a few hubs that control the network. It is worth noting that many real-world networks, including the world wide web linking structure, are known to show a right-skewed distribution (often a power law $P(k) \propto k^{-\gamma}$ with $2 < \gamma < 3$).

5.3.3.2 Degree-Degree Correlation

Another metric used is the degree correlation. It is equal to the average nearest-neighbour degree, k^{nn}, as a function of k:

$$k^{nn}(k) = \sum_{k'=0}^{\infty} k' p(k'|k), \qquad (5.22)$$

where $p(k'|k)$ is the fraction of edges that are attached to a vertex of degree k whose other ends are attached to vertex of degree k'. Thus, $k^{nn}(k)$ is the mean degree of the vertices we find by following a link emanating from a vertex of degree k.

A closely related concept is the degree–degree correlation coefficient, also known as *assortative mixing*, which is the Pearson r correlation coefficient for degrees of vertices at either end of a link. A monotonically increasing (decreasing) k^{nn} means that high-degree vertices are connected to other high-degree (low-degree) vertices, resulting in a positive (negative) value of r [10]. In recommender systems, it measures to which extent nodes are connected preferentially to other nodes with similar characteristics.

5.3.3.3 Mixing Patterns

We can generalise the vertex assortative mixing to any network pattern. Assortative mixing has an impact on the structural properties of the network. Mixing by a discrete characteristic of the network (e.g. race, language, or age in social networks) tend to separate the network into different communities. In social networks, this is also known as *homophily*.

We use the formula defined in [11] to compute mixing patterns for discrete attributes. Let E be an $N \times N$ matrix, where E_{ij} contains the number of edges in the network that connect a vertex of type i to one of type j ($E_{ij} = E_{ji}$ in undirected networks). The normalised mixing matrix is defined as:

$$\mathbf{e} = \frac{E}{\| E \|} \qquad (5.23)$$

where $\| x \|$ means the sum of all elements in the matrix x. Mixing characteristics is measured in the normalised matrix $\mathbf{e}$. Matrix $\mathbf{e}$ satisfies the following sum rules:

$$\sum_{ij} e_{ij} = 1, \tag{5.24}$$

$$\sum_{j} e_{ij} = a_i, \tag{5.25}$$

$$\sum_{i} e_{ij} = b_j, \tag{5.26}$$

where a_i and b_i are the fraction of each type of an end of an edge that is attached to nodes of type i. The assortative mixing coefficient r is defined as:

$$r = \frac{\sum_i e_{ii} - \sum_i a_i b_i}{1 - \sum_i a_i b_i} = \frac{Tr(\mathbf{e}) - \| \mathbf{e}^2 \|}{1 - \| \mathbf{e}^2 \|} \tag{5.27}$$

This quantity equals to 0 in a randomly mixed network, and 1 in a perfectly mixed network. Dissassortative networks have a negative r value, whilst assortative networks have a positive one.

5.3.4 Clustering

5.3.4.1 Local Clustering Coefficient

The local clustering coefficient, C_i, of a node i represents the probability of its neighbours to be connected within each other.

$$C_i = \frac{2|E_i|}{k_i(k_i - 1)}, \tag{5.28}$$

where E_i is the set of existing edges that are direct neighbours of i, and k_i the degree of i. C_i denotes, then, the portion of actual edges of i from the potential number of total edges. $\langle C \rangle$ is defined as the average over the local measure C_i [12]:

$$\langle C \rangle = \frac{1}{n} \sum_{i=1}^{n} C_i \tag{5.29}$$

5.3.4.2 Global Clustering Coefficient

The global clustering coefficient is a sign of how cliquish (tightly knit) a network is. It estimates the conditional probability that two neighbouring vertices of a given vertex are neighbours themselves. The global clustering coefficient, C, It is quantified by the abundance of triangles in a network, where a triangle is formed when three vertices are all linked to one another.

$$C = \frac{3 \times \text{number of triangles}}{\text{number of connected triples}}.$$ (5.30)

Here, a connected triple means a pair of vertices connected via another vertex. Since a triangle contains three triples, C is equal to the probability that two neighbours of a vertex are connected as well. For random graphs, the clustering coefficient is defined as $C_r \sim \langle k \rangle / N$. Typically, real networks have a higher clustering coefficient than C_r.

Some real-world networks are known to show a behaviour of $C(k) \propto k^{-1}$, usually attributed to the hierarchical nature of the network [13]. This behaviour has been found in metabolic networks, as well as in the WWW, and movie actor networks [14]. The reasons for modular organisation in these networks relate, respectively, to the function in metabolic interaction networks, the topology of Internet, and the social activities in social networks.

5.3.5 Centrality

Centrality, as its name suggests, is a concept that differentiates vertices according to how influential they are in a network. Given the inhomogeneity of link patterns around vertices in a complex network, we could certainly imagine that the position and roles of vertices will vary significantly from one vertex to another.

5.3.5.1 Degree

Degree centrality is defined as the number of links incident upon a node. It is reasonable to assume that items with particularly many acquaintances can be looked as being important figures. However, degree is primarily local in scope (e.g. talking loudly does not mean that you are affecting others more effectively than somebody who speaks quietly but very eloquently).

5.3.5.2 Closeness

Closeness centrality is defined as the mean geodesic distance between a vertex v and all other vertices reachable from it. Those nodes that tend to have short geodesic distances to other vertices within the graph have higher closeness. Closeness can be regarded as a measure of how long it will take information to spread from a given vertex to other reachable vertices in the network [15].

$$C_C(v) = \frac{\sum_{t \in V \setminus v} d(v,t)}{N-1} \qquad (5.31)$$

where $N \geq 2$ is the size of the graph component reachable from node v.

A complementary way to define closeness centrality is the reciprocal of the sum of geodesic distances to all other vertices of v [16].

$$C_C(v) = \frac{1}{\sum_{t \in V \setminus v} d(v,t)} \qquad (5.32)$$

5.3.5.3 Betweenness

Betweenness Freeman centrality measures whether a central vertex will act as a relay of information between vertices, a role endowed thanks to being on a geodesic between vertices (hence the name betweenness) [17].

The definition of *Freeman (betweenness) centrality* C_B of a vertex v is defined as:

$$C_B(v) = \frac{1}{2} \sum_{i,j} \frac{g_{ivj}}{g_{ij}}, \qquad (5.33)$$

where g_{ij} is the total number of geodesics between vertices i and j, and g_{ivj} is the number of the ones that pass through the vertex v.

5.3.6 Limitations

The main limitation of the network-centric approach is that users remain outside the evaluation process. There is no user intervention, not even the information of a user profile is taken into account in the evaluation. The main drawbacks of network-centric approach are:

- Accuracy of the recommendations cannot be measured. In the network-centric approach there is no way to evaluate *how well* the algorithm is predicting the items already in a user's profile.
- Neither *transparency* (explainability) nor *trustworthiness* (confidence) of the recommendations can be measured.
- The *perceived quality* (i.e. usefulness and effectiveness) of the recommendations cannot be measured. The only way to solve this limitation is by letting users to step in the evaluation process.

5.3.7 Related Work in Music Information Retrieval

During the last few years, complex network analysis has been applied to music information retrieval, and music recommendation in particular. In [18], we compared different music recommendation algorithms based on the network topology. The results show that social based recommenders present a scale-free network topology, whereas human expert-based controlled networks does not.

An empirical study of the evolution of a social network constructed under the influence of musical tastes, based on playlist co-occurrence, appears in Buldu et al. [19]. The analysis of collaboration among contemporary musicians, in which two musicians are connected if they have performed in or produced an album together, appears in [20]. Anglade et al. present a user clustering algorithm that exploits the topology of a user-based similarity network [21].

Aucouturier presents in [22] a network of similar songs based on timbre similarity. Interestingly enough, the network is scale-free, thus a few songs appear in almost any list of similar tracks. This has some problems when generating automatic playlists. Jacobson and Sandler [23] present an analysis of the Myspace social network, and conclude that artists tend to form on-line communities with artists of the same musical genre.

In [24], Lambiotte and Ausloos present a method for clustering genres, by analysing correlations between them. The analysis is based on the users' listening habits, gathered from *last.fm*. From the $\langle user, artist, plays \rangle$ triples the authors compute genre similarity based on the percolation idea in complex networks, and also visualise a music genre cartography, using a tree representation.

5.4 User-Centric Evaluation

User-centric evaluation aims at measuring the user's perceived quality and usefulness of the recommendations. In this case, the evaluation requires the user intervention to provide feedback of the provided recommendations—via a survey, or gaterhing information from the user activity in the system.

User-centric evaluation copes with the limitations of both system- and network-centric approaches. Figure 5.3 depicts this method, we named *user–centric evaluation* with feedback. Two important limitations of system- and network-centric approaches are the impossibility to evaluate the novelty and the perceived quality of the recommendations. User-centric allows us to evaluate these two elements. The main difference with a system-centric approach is that user-centric expands the evaluation dataset to those items that the user has not yet seen (i.e. rated, purchased, previewed, etc.).

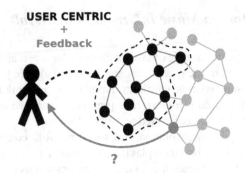

Fig. 5.3 User-centric evaluation, including feedback about the received recommendations.

5.4.1 Gathering Feedback

In the user-centric approach, the recommender system presents relevant items (from outside the user's dataset), and asks user for feedback. Feedback gathering can be done in two ways: implicitly or explicitly. Measuring implicit feedback includes, for instance, the time spent in the item's webpage, purchasing or not the item, previewing it, etc. Explicit feedback is based on two related questions; (i) whether the user already knew the item (novelty), and (ii) whether she likes it or not (perceived quality). Obviously, it requires an extra effort from the users, but at the same time it provides unequivocal information about the intended dimensions (which in the case of implicit measures could be ambiguous or inaccurate).

5.4.1.1 Perceived Quality

The easiest way to measure the perceived quality of the recommended items is by explicitly asking to the users. Users must examine the recommended items and validate, to some extent, whether they like the items or not [2]. In this sense, a user needs the maximum information about the item (e.g. metadata information, a preview, etc.), and the reasons why the item was recommended, if possible. Then, the user has to rate the quality of each recommended item (e.g. in a rating scale of [1..5]), or the quality of the list as a whole. Last but not least, the user should be able to select those attributes of the item that makes her feel that the novel item is relevant to her taste.

5.4.1.2 Novelty

To evaluate novel items we need, again, to ask to the users whether they recognise the predicted item or not. Users have to examine the list of recommended items and express, for each item, whether she previously knew the item or not.

Combining both aspects, perceived quality and novelty, allows the system to infer whether a user likes to receive and discover unknown items, or in contrast, she prefers to get more conservative and familiar recommendations. Adding the transparency (explainability) in the recommendations, the user can perceive the new items as of higher quality, as the system can give an explanation of why this unknown item was recommended to the user. All in all, the user's intentions with regard novelty detection depends on the context and the recommendation domain. Furthermore, it is expected that the intentions change over time. For instance, a user is sometimes open to discovering new artists and songs, while sometimes she just wants to listen to her favourites. Detecting these modes and acting accordingly would increase user's satisfaction with the system.

5.4.1.3 A/B Testing

Another approach to gather feedback from the users is via an A/B test. In A/B testing, the system unleash two different versions of an algorithm (or two completely different algorithms), to evaluate which one performs the best. The performance is measured by the impact the new algorithm (say A) has on the visitors' behaviour, compared with the baseline algorithm (B). A/B testing became very popular on the Web, because it is easy to create different webpage versions, and show them to visitors. One of the first successful examples that used A/B test was *Amazon.com*. When they saw the results, they decided to show recommendations (similar products) in the product page.[1]

In A/B testing, the evaluation is performed by only changing a few aspects between the two versions. Once a baseline is established, the system starts optimising the algorithm by making one change at a time, and evaluating the results and impact with real visitors of the page. A/B testing uses *between subjects* evaluation. That is, the system splits the users in two groups. Each group only evaluates one approach (or algorithm), but not the other. Contrastingly, in a *within subjects* evaluation each user evaluates all the possible approaches (or algorithms) instead of only one.

5.4.2 Limitations

The main limitation of the user-centric approach is the need of user intervention in the evaluation process. Gathering feedback from the user can be tedious for some

---

[1] http://glinden.blogspot.com/2006/04/early-amazon-shopping-cart.html

users (filling surveys, rating items, providing feedback, etc.). In this sense, the system should ease and minimise the user intervention, using (whenever is possible) an unintrusive way. On the other hand, the main limitations from the two previous approaches (perceived quality and novelty detection) are solved in this approach.

5.5 Summary

We classify the evaluation of recommender algorithms in: system-, network-, and user-centric approaches. System-centric evaluation measures how accurately the recommender system can predict the actual values that users have previously assigned. Network-centric evaluation aims at measuring the topology of the item (or user) network similarity, and it uses metrics from complex network analysis. Finally, user-centric evaluation focuses on the user's perceived quality and usefulness of the recommendations. Combining the three methods we can cover all the facets of a recommender algorithm; the system-centric approach evaluates the performance accuracy of the algorithm, the network-centric approach analyses the structure of the similarity network, and with the inclusion of the user intervention we can measure the satisfaction about the recommendations they receive. Figure 5.4 depicts this idea. We can see that, when using the three evaluation approaches, all the components are evaluated—algorithm accuracy, similarity network analysis, and feedback from users.

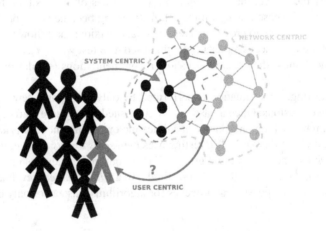

Fig. 5.4 *System-, network-,* and *user-centric* evaluation methods. Combining the three methods we can cover all the facets when evaluating a recommendation system.

Last but not least, Table 5.2 summarises the limitations of each approach. The table presents some of the factors that affect the recommendations, and whether the approach can evaluate it or not. Applying the three evaluation approaches, we can assess all the facets of a recommender system, and also cope with the limitations of each evaluation approach.

	Accuracy	Coverage	Novelty	Diversity	Transp.	Quality
System-centric	✓	✗	✗	✓	✗	✗
Network-centric	✗	✓	✓	✓	✗	✗
User-centric	✗	✗	✓	✓	✓	✓

Table 5.2 A summary of the evaluation methods. It shows the factors that affect the recommendations, and whether the approach can evaluate it or not.

5.5.1 Links with the Following Chapters

In this chapter we have presented the three methods to evaluate recommender algorithms. In the following two chapters we apply the metrics in real recommendation datasets. The evaluation based on network-centric is presented in Chap. 6. Then, user-centric evaluation is presented in Chap. 7. In the remaining of the book, we do not present any results using system-centric metrics.

References

1. J. S. Breese, D. Heckerman, and C. Kadie, "Empirical analysis of predictive algorithms for collaborative filtering," tech. rep., Microsoft Research, 1998.
2. J. L. Herlocker, J. A. Konstan, L. G. Terveen, and J. T. Riedl, "Evaluating collaborative filtering recommender systems," *ACM Transaction on Information Systems*, vol. 22, no. 1, pp. 5–53, 2004.
3. G. Shani and A. Gunawardana, "Evaluating recommender systems," tech. rep., Microsoft Research, MSR-TR-2009–159, November 2009.
4. M. Balabanovic and Y. Shoham, "Fab: Content-based, collaborative recommendation," *Communications of the ACM*, vol. 40, pp. 66–72, 1997.
5. Y. Y. Yao, "Measuring retrieval effectiveness based on user preference of documents," *Journal of the American Society for Information Science*, vol. 46, no. 2, pp. 133–145, 1995.
6. G. Adomavicius and A. Tuzhilin, "Toward the next generation of recommender systems: A survey of the state-of-the-art and possible extensions," *IEEE Transactions on Knowledge and Data Engineering*, vol. 17, no. 6, pp. 734–749, 2005.
7. S. M. McNee, J. Riedl, and J. A. Konstan, "Being accurate is not enough: How accuracy metrics have hurt recommender systems," in *Computer Human Interaction. Human factors in computing systems*, (New York, NY), pp. 1097–1101, ACM, 2006.
8. P. Erdös and A. Réyi, "On random graphs," *Science*, vol. 6, no. 290, pp. 290–298, 1959.
9. A. L. Barabási and R. Albert, "Emergence of scaling in random networks," *Science*, vol. 286, pp. 509–512, October 1999.
10. M. E. J. Newman, "Assortative mixing in networks," *Physical Review Letters*, vol. 89, no. 20, 2002.

11. M. E. J. Newman, "Mixing patterns in networks," *Physical Review E*, vol. 67, 2003.
12. D. J. Watts and S. H. Strogatz, "Collective dynamics of 'small-world' networks," *Nature*, vol. 393, pp. 440–442, June 1998.
13. E. Ravasz and A. L. Barabási, "Hierarchical organization in complex networks," *Physical Review. E, Statistical, Nonlinear, and Soft Matter Physics*, vol. 67, February 2003.
14. E. Ravasz, A. L. Somera, D. A. Mongru, Z. N. Oltvai, and A. L. Barabási, "Hierarchical organization of modularity in metabolic networks," *Science*, vol. 297, no. 5586, pp. 1551–1555, 2002.
15. M. Newman, "A measure of betweenness centrality based on random walks," *Social Networks*, vol. 27, pp. 39–54, January 2005.
16. G. Sabidussi, "The centrality index of a graph," *Psychometrika*, vol. 31, pp. 581–603, December 1966.
17. L. C. Freeman, "Centrality in social networks: Conceptual clarification," *Social Networks*, vol. 1, no. 3, pp. 215–239, 1979.
18. P. Cano, O. Celma, M. Koppenberger, and J. Martin-Buldú, "Topology of music recommendation networks," *Chaos: An Interdisciplinary Journal of Nonlinear Science*, vol. 16, no. 013107, 2006.
19. J. Martin-Buldú, P. Cano, M. Koppenberger, J. Almendral, and S. Boccaletti, "The complex network of musical tastes," *New Journal of Physics*, vol. 9, no. 172, 2007.
20. J. Park, O. Celma, M. Koppenberger, P. Cano, and J. Martin-Buldú, "The social network of contemporary popular musicians," *International Journal of Bifurcation and Chaos*, vol. 17, no. 7, pp. 2281–2288, 2007.
21. A. Anglade, M. Tiemann, and F. Vignoli, "Complex-network theoretic clustering for identifying groups of similar listeners in p2p systems," in *Proceedings of the ACM conference on Recommender systems*, (Minneapolis, MN), pp. 41–48, ACM, 2007.
22. J.-J. Aucouturier and F. Pachet, "A scale-free distribution of false positives for a large class of audio similarity measures," *Pattern Recognition*, vol. 41, no. 1, pp. 272–284, 2008.
23. K. Jacobson and M. Sandler, "Musically meaningful or just noise? an analysis of on-line artist networks," in *Proceedings of the 6th International Symposium on Computer Music Modeling and Retrieval*, (Copenhagen, Denmark), 2008.
24. R. Lambiotte and M. Ausloos, "Uncovering collective listening habits and music genres in bipartite networks," *Physical Review E*, vol. 72, 2005.

Chapter 6
Network-Centric Evaluation

In this chapter we present the network-centric evaluation approach. This method analyses the similarity network, created using any recommendation algorithm. Network-centric evaluation uses complex networks analysis to characterise the item collection. Also, we can combine the results from the network analysis with the popularity of the items, using the Long Tail model.

We perform several experiments in the music recommendation field. The first experiment aims at evaluating the popularity effect using three music artists recommendation approaches: collaborative filtering (CF), content-based audio similarity (CB), and human expert-based resemblance. The second experiment compare two user networks created by CF and CB derived from the users' listening habits. In all the three experiments, we measure the popularity effect by contrasting the properties from the network with the Long Tail information (e.g. are the hubs in the recommendation network the most popular items? Or, are the most popular items connected with other popular items?).

6.1 Network Analysis and the Long Tail Model

Figure 6.1 presents the framework for the network-centric evaluation. It includes the similarity network and the Long Tail of item popularity. This approach combines the analysis of the similarity network with the Long Tail of popularity.

Once each item in the recommendation network is located in the head, mid, or tail part (see Sec. 4.3.2), the next step is to combine the similarity network with the Long Tail information. Two main analysis are performed: first, we measure the similarity among the items in each part of the curve. That is, for each item that belongs to the head part, compute the percentage of similar items that are located in the head, mid and tail part (similarly, for the items in the mid and tail part). This measures whether the most popular items are connected with other popular items, and vice versa. Second, we measure the correlation between an item's rank in the

Ò. Celma, *Music Recommendation and Discovery*,
DOI 10.1007/978-3-642-13287-2_6, © Springer-Verlag Berlin Heidelberg 2010

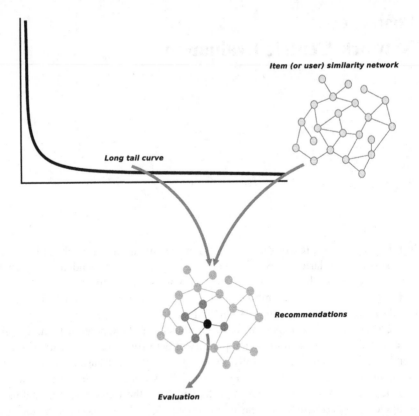

Fig. 6.1 General framework for the network-centric evaluation. The network-centric approach determines the underlying topology of the similarity network, and combines this information with the Long Tail of popularity.

Long Tail and its indegree. This measure allows us to detect whether the hubs in the network are also the most popular items.

Section 6.2 presents the experiments about the popularity effect in three different music artists recommendation algorithms: collaborative filtering (CF) from *last.fm*, content-based audio filtering (CB), and expert-based recommendations (EX) from *Allmusic.com* (*AMG*) musicologists. Then, Sec. 6.3 compares two user similarity networks created using collaborative filtering (CF) again from *last.fm*, and a user similarity network derived from the users' listening habits. In this case, we use content-based audio similarity (CB) to create the links among users.

6.2 Artist Network Analysis

We aim to evaluate three artist similarity networks: collaborative filtering (CF), content-based audio similarity (CB), and human expert-based resemblance. Also, we analyse the popularity effect for each recommendation network. We measure the popularity effect by contrasting the properties from the network with the Long Tail information of the catalog.

6.2.1 Datasets

6.2.1.1 Social-Based, Collaborative Filtering Network

Artist similarity is gathered from *last.fm*, using Audioscrobbler web services,[1] and selecting the top-20 similar artists. *Last.fm* has a strong social component, and their recommendations are based on a combination of an item-based collaborative filtering, plus the information derived from social tagging. We denote this network as *CF*.

6.2.1.2 Human Expert-Based Network

We have gathered human-based expert recommendations from *All Music Guide* (*AMG*).[2] AMG makes use of professional editors to interconnect artists, according to several aspects, such as: *influenced by, followers of, similar artists, performed songs by*, etc. In order to create an homogeneous network, we only use the *similar artists* links. We denote this network as *EX*.

Table 6.1 shows the number of nodes and edges, for each network.

	Number of artists	Number of relations
Last.fm social filtering (CF)	122,801	1,735,179
Allmusic.com expert-based (EX)	74,494	407,483
Content-based (CB)	59,583	1,179,743

Table 6.1 Datasets for the artist similarity networks.

---

[1] http://www.audioscrobbler.net/data/webservices/

[2] http://www.allmusic.com

6.2.1.3 Content-Based Network

To compute artist similarity in the CB network, we apply content-based audio analysis in an in-house music collection ($\mathcal{T}$) of 1.3 Million tracks of 30 s samples. Distance between tracks, $d(x,y)$, is based on the Euclidean distance over a reduced space using Principal Component Analysis (PCA). The audio features used include not only timbral features (e.g. Mel frequency cepstral coefficients), but musical descriptors related to rhythm (e.g. beats per minute, perceptual speed, binary/ternary metric), and tonality (e.g chroma features, key and mode), among others [1]. Preliminary steps to compute the Euclidean distance are: (i) audio descriptor normalisation in the $[0, 1]$ interval, and (ii) applying PCA to reduce the audio descriptors space to 25 dimensions.

Then, to compute artist similarity we used the most representative tracks, $\mathcal{T}_a$, of an artist a, with a maximum of 100 tracks per artist. For each track, $t_i \in \mathcal{T}_a$, we obtain the most similar tracks (excluding those from artist a):

$$sim(t_i) = \underset{\forall t \in \mathcal{T}}{\operatorname{argmin}} \, (d(t_i, t)), \qquad (6.1)$$

and get the artists' names, $\mathcal{A}_{sim(t_i)}$, of the similar tracks. The list of (top-20) similar artists of a is composed by all $\mathcal{A}_{sim(t_i)}$, ranked by frequency and weighted by the audio similarity distance:

$$similar\_artists(a) = \bigcup \mathcal{A}_{sim(t_i)}, \forall t_i \in \mathcal{T}_a \qquad (6.2)$$

6.2.2 Network Analysis

6.2.2.1 Small World Navigation

Table 6.2 shows the network properties of the three datasets. All the networks exhibit the *small-world* phenomena [2]. Each network has a small directed shortest path $\langle d_d \rangle$ comparable to that of their respective random network. Also all the clustering coefficients, C, are significantly higher than the equivalent random networks C_r. This is an important property, because recommender systems can be structurally optimised to allow surfing to any part of a music collection with a few of mouse clicks, and so that they are easy to navigate using only local information [3, 4].

The human-expert network has a giant component, SGC, smaller than CF and CB networks. More than 4% of the artists in the human-expert network are isolated, and cannot be reached from the rest. This has strong consequences concerning the coverage of the recommendations and network navigation.

Property	CF (*Last.fm*)	EX (*AMG*)	CB
N	122,801	74,494	59,583
$\langle k \rangle$	14.13	5.47	19.80
$\langle d_d \rangle (\langle d_r \rangle)$	5.64 (4.42)	5.92 (6.60)	4.48 (4.30)
D	10	9	7
SGC	99.53%	95.80%	99.97%
γ_{in}	2.31($\pm$0.22)	NA (log-normal)	1.61($\pm$0.07)
r	0.92	0.14	0.17
$C (C_r)$	0.230 (0.0001)	0.027 (0.00007)	0.025 (0.0002)

Table 6.2 Artist recommendation network properties for *last.fm* collaborative filtering (CF), content-based audio filtering (CB), and *Allmusic.com* (*AMG*) expert-based (EX) networks. N is the number of nodes, and $\langle k \rangle$ the mean degree, $\langle d_d \rangle$ is the avg. shortest directed path, and $\langle d_r \rangle$ the equivalent for a random network of size N, D is the diameter of the (undirected) network. SGC is the size (percentage of nodes) of the strong giant component for the undirected network, γ_{in} is the power-law exponent of the cumulative indegree distribution, r is the indegree–indegree Pearson correlation coefficient (assortative mixing), C is the clustering coefficient for the undirected network, and C_r for the equivalent random network.

6.2.2.2 Clustering Coefficient

The clustering coefficient for the CF network is significantly higher than that of the CB or EX networks ($C_{CF} = 0.230$). This means, given an artist a, the neighbours of a are also connected with each other with a probability of 0.230. For instance, *U2*'s list of similar artists includes *INXS* and *Crowded House*, and these two bands are also connected, forming a triangle with *U2*. This has an impact on the navigation of the network, as one might get stuck in a small cluster.

6.2.2.3 Indegree Distribution

The shape of the (complementary) cumulative indegree distribution informs us about the topology of the recommendation network (random, or scale-free). We follow the steps defined in Sec. 4.4 to decide whether or not the indegree distribution follows a power-law (and, thus, it is a scale-free network).

	power-law	power-law + cut-off			log-normal		support for
	p	LLR	p	x_{cutoff}	LLR	p	power-law
CF	0.9	−165.48	0.00	≈ 102	−25.15	0.00	with exp. decay cut-off
Expert	0.43	−41.05	0.00	≈ 66	−5.86	0.00	moderate, with cut-off
CB	0.12	−905.96	0.00	≈ 326	−99.68	0.00	moderate, with cut-off

Table 6.3 Model selection for the indegree distribution of the three artist networks. For each network we give a *p-value* for the fit to the power-law model (first column). The first *p-value* equals to the Kolmogorov–Smirnov D statistic (see Eq. 4.8). We also present the likelihood ratios for the alternative distributions (power-law with an exponential cut-off, and log-normal), and the *p-values* for the significance of each of the likelihood ratio tests (LLR).

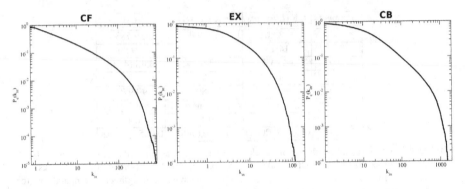

Fig. 6.2 Cumulative indegree distribution for the three artist networks.

Table 6.3 presents the model selection for the indegree distribution. For each network we give a *p-value* for the fit to the power-law model (first column). A higher *p-value* means that the distribution is likely to follow a power-law. In Table 6.3, we also present the likelihood ratios for the alternative distributions (power-law with an exponential cut-off, and log-normal), and the *p-values* for the significance of each of the likelihood ratio tests. In this case, a *p-value* close to zero means that the alternative distribution can also fit the distribution. In all the three networks, the distribution can be fitted using either a power-law with an exponential decay, or a log-normal. For the log-normal, non-nested alternative, we give the normalised log likelihood ratio $R/\sqrt{n}\sigma$, as [5]. For the power law with an exponential cut-off, a nested distribution, we give the actual log likelihood ratio. The final column of the table lists our judgement of the statistical support for the power-law hypothesis for each artist network.

The best fit for the CF network (according to the log-likelihood[3]) is obtained with a power-law with an exponential decay (starting at $x_{cutoff} \approx 102$), $x^{-2.31}e^{-7x}$. In the expert-based network, the best fit (with a log-likelihood of 581.67) is obtained with a log-normal distribution, $\frac{1}{x}e^{-\frac{(\ln(x)-\mu)^2}{2\sigma^2}}$, with parameters mean of log $\mu = 7.36$, and standard deviation of log, $\sigma = 3.58$. Finally, the CB network follows a moderate a power-law with an exponential decay, $x^{-1.61}e^{-7.19x}$ ($x_{cutoff} \approx 326$). Yet, in this case the log-normal can be considered as good as the power-law distribution with cut-off.

Figure 6.2 shows the cumulative indegree distribution for each network. EX follows a log-normal distribution, whereas CF and CB follow a power law with an exponential decay (cut–off). CF has a power-law exponent, $\gamma = 2.31$, similar to those detected in many *scale free* networks, including the world wide web linking structure [6]. These networks are known to show a right-skewed power law distribution, $P(k) \propto k^{-\gamma}$ with $2 < \gamma < 3$, relying on a small subset of hubs that control the network [7].

---

[3] Not to be confused with the Log-likelihood **ratio** (LLR), that we use to compare two distributions.

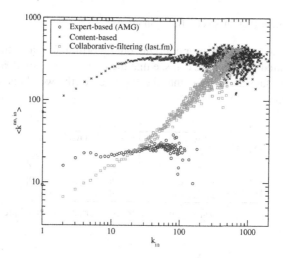

Fig. 6.3 Indegree–indegree correlation (assortative mixing) for the three artist recommendation networks: collaborative filtering (CF) from *last.fm*, Content-based (CB), and *Allmusic.com* experts. CF clearly presents the assortative mixing phenomenon ($r_{CF} = 0.92$). Neither CB nor expert-based present any correlation ($r_{CB} = 0.14$, $r_{Expert} = 0.17$).

6.2.2.4 Assortative Mixing

Another difference in the three networks is the assortative mixing, or indegree–indegree correlation. Figure 6.3 shows the correlation for each network. The CF network presents a high assortative mixing ($r = 0.92$). That means that the most connected artists are prone to be similar to other top connected artists. Neither CB nor EX present indegree–indegree correlation, thus artists are connected independently of their inherent properties.

6.2.2.5 Mixing by Genre

We are also interested in the assortative mixing of the network, according to the musical genre. E.g. do similar artists tend to belong to the same genre? To do this, we gather the artists' tags from *last.fm*, and filter those tags that do not refer to a genre. To match the tags with a predefined list of 13 seed genres, we follow the approach presented in [8]. Listing 6.1 shows an snippet of the *last.fm* normalised tags for *Bruce Springsteen* (tag weight ranges [1..100]):

```
Bruce Springsteen       classic rock    100
Bruce Springsteen       rock            95
Bruce Springsteen       pop             80
Bruce Springsteen       80s             72
Bruce Springsteen       classic         50
Bruce Springsteen       folk-rock       25
...
```

Listing 6.1 Snippet of Last.fm tags for Bruce Springsteen.

Table 6.4 shows the result after applying our algorithm to match the genres from the list of weighted tags [8]. We can see that the tag 80s is filtered, and classic rock and rock tags are merged into the *Rock* genre (the weight is the sum of the two tags' weights).

Tag	Matched genre	Weight
classic rock, rock	Rock	195
pop	Pop	80
classic	Classical	50
folk-rock	Folk	25

Table 6.4 Assigned genres for *Bruce Springsteen* from the artist's tag cloud presented in Listing 6.1.

Once we get the matched genres for all the artists, we can analyse whether similar artists tend to belong to the same (or a semantically close) genre. Mixing correlation by genre coefficient r is computed using Eq. (5.27), over the normalised correlation matrix **e** (see Eq. 5.23). We create the correlation matrix **e** for the three networks following three steps:

1. For each artist a_i, get the list of weighted genres $\mathcal{G}_{a_i}$, as well as the list of genres from the similar artists of a_i, $\mathcal{G}_{sim(a_i)}$.
2. Create the correlation matrix **E**. For each genre $g_{a_i} \in \mathcal{G}_{a_i}$, and $g_j \in \mathcal{G}_{Sim(a_i)}$, increment $\mathbf{E}_{g_{a_i},g_j}$ combining the artist similarity value, $similarity(a_i,a_j)$, for artists $a_j \in Sim(a_i)$, with the sum of the two genres' weights.
   $\mathbf{E}_{g_{a_i},g_j} = \mathbf{E}_{g_{a_i},g_j} + similarity(a_i,a_j) \cdot (g_{a_i} + g_j)$
3. Create the normalised correlation matrix **e** from **E**, using Eq. (5.23), and normalising it with $\sum_{ij} e_{ij} = 100$.

Tables 6.5, 6.6, and 6.7 present the matrices **e** for the CF, EX and CB networks, (bold denotes the highest values) respectively. Then, Table 6.8 shows the assortative mixing coefficient r for each network, computed over **e** (using Eq. 5.23). The higher r coefficient is found in the human expert network, $r_{EX} = 0.411$. According to human experts, then, artist genre is a relevant factor to determine artist similarity. As expected, the content-based network does not present mixing by genre ($r_{CB} = 0.089$). Our results are aligned with the findings in [9]. They use the *Myspace.com* network of artists' friends, and set only one genre label per artist. The mixing by genre coefficient value they obtain is $r = 0.350$. Therefore, *Myspace* artists prefer to maintain friendship links with other artists in the same genre.

In our three artist networks, *metal, pop, punk* and *rock* genres accumulate more than 50% of the fraction of links (see a_i, last column of the tables). So, the three networks are biased towards these few genres, which have a big impact in the similarity network. This bias concords with the type of users in the *last.fm* community, and the

	Blues	Classic	Ctry	Elec	Folk	Jazz	Metal	Pop	Punk	Rock	Rap	Regg	Soul	a_i
Blues	**1.09**	0.01	0.27	0.05	0.11	0.18	0.12	0.36	0.08	0.35	0.02	0.02	0.07	*2.74*
Classical	0.01	0.07	0.01	0.06	0.02	0.04	0.08	**0.15**	0.07	**0.15**	0.03	0.00	0.01	*0.71*
Country	0.47	0.02	**2.31**	0.08	0.22	0.12	0.06	0.36	0.10	0.37	0.04	0.02	0.04	*4.22*
Electronic	0.03	0.03	0.03	**4.17**	0.07	0.13	0.48	1.27	0.52	1.14	0.3	0.06	0.05	*8.28*
Folk	0.07	0.01	0.11	0.08	**0.59**	0.04	0.10	0.29	0.08	0.33	0.02	0.01	0.01	*1.73*
Jazz	0.19	0.03	0.11	0.29	0.07	**1.30**	0.20	0.46	0.20	0.44	0.11	0.04	0.10	*3.53*
Metal	0.09	0.05	0.02	0.54	0.10	0.12	**8.74**	1.81	1.20	2.95	0.20	0.04	0.01	*15.88*
Pop	0.23	0.07	0.13	1.15	0.26	0.18	1.54	**7.28**	1.46	3.46	0.31	0.06	0.06	*16.2*
Punk	0.06	0.05	0.04	0.57	0.09	0.12	1.37	1.85	**5.29**	1.80	0.24	0.06	0.03	*11.58*
Rock	0.34	0.09	0.26	1.83	0.45	0.34	3.33	4.71	2.23	**12.06**	0.52	0.16	0.20	*26.52*
Rap	0.02	0.01	0.01	0.40	0.02	0.08	0.22	0.45	0.26	0.42	**2.50**	0.04	0.04	*4.46*
Reggae	0.02	0.01	0.02	0.14	0.02	0.07	0.08	0.26	0.16	0.25	0.09	**2.23**	0.04	*3.38*
Soul	0.04	0.00	0.02	0.05	0.01	0.05	0.01	0.09	0.04	0.12	0.04	0.01	**0.28**	*0.76*
b_j	*2.66*	*0.44*	*3.34*	*9.42*	*2.03*	*2.77*	*16.34*	*19.33*	*11.69*	*23.86*	*4.42*	*2.76*	*0.93*	*100*

Table 6.5 Normalised mixing matrix $\mathbf{e}^{CF}$ for the *last.fm* network.

	Blues	Classic	Ctry	Elec	Folk	Jazz	Metal	Pop	Punk	Rock	Rap	Regg	Soul	a_i
Blues	**2.75**	0.06	0.60	0.03	0.18	0.67	0.18	0.40	0.08	0.80	0.01	0.02	0.09	*5.88*
Classical	0.03	0.20	0.03	0.05	0.03	0.21	0.06	0.12	0.06	**0.35**	0.02	0.01	0.01	*1.18*
Country	0.84	0.05	**6.07**	0.05	0.45	0.41	0.04	0.32	0.05	0.74	0.02	0.02	0.04	*9.09*
Electronic	0.04	0.07	0.05	**1.66**	0.05	0.16	0.18	0.41	0.17	0.74	0.15	0.03	0.03	*3.75*
Folk	0.15	0.03	0.31	0.05	**0.99**	0.09	0.05	0.20	0.04	0.52	0.01	0.01	0.01	*2.46*
Jazz	0.70	0.33	0.28	0.18	0.10	**11.71**	0.10	0.27	0.10	1.14	0.08	0.05	0.12	*15.17*
Metal	0.18	0.09	0.04	0.19	0.08	0.09	**4.17**	1.28	0.63	2.84	0.12	0.04	0.02	*9.78*
Pop	0.33	0.13	0.19	0.38	0.22	0.19	1.06	**3.48**	0.56	3.39	0.17	0.05	0.05	*10.21*
Punk	0.07	0.07	0.05	0.22	0.05	0.10	0.68	0.87	**1.74**	1.61	0.12	0.05	0.03	*5.66*
Rock	0.79	0.44	0.72	0.60	0.48	1.34	2.10	2.71	0.88	**20.35**	0.24	0.18	0.19	*31.01*
Rap	0.01	0.02	0.02	0.16	0.01	0.06	0.09	0.18	0.09	0.30	**1.32**	0.02	0.04	*2.33*
Reggae	0.03	0.01	0.02	0.06	0.02	0.06	0.05	0.13	0.07	0.25	0.03	**2.07**	0.03	*2.82*
Soul	0.06	0.01	0.02	0.03	0.01	0.08	0.02	0.07	0.02	**0.22**	0.03	0.01	0.07	*0.66*
b_j	*5.98*	*1.53*	*8.41*	*3.65*	*2.66*	*15.18*	*8.78*	*10.46*	*4.49*	*33.24*	*2.32*	*2.59*	*0.72*	*100*

Table 6.6 Normalised mixing matrix $\mathbf{e}^{EX}$ for the AMG human-expert network.

tags they apply the most. The EX and CB networks have more *country* artists than the CF network artists. Also, in the expert network there is a lot of *jazz* artists. Additionally, in the three networks there is an underrepresentation of the *classical*, *folk* and *soul* artists. The reality is that a recommender system has to deal with biased collections, and make the best out of it.

In terms of genre cohesion, *classical* is always "misclassified" as *pop/rock*. In our case, the problem with the *classical* genre is that some non-classical music artists are tagged as *classic*. Our algorithm matches this tag with the seed genre *Classical* (see the Bruce Springsteen example in Table 6.4). Actually, if we remove the classical genre from the list of 13 genres, the *r* correlation coefficient increases by 0.1, in the CF and EX networks.

	Blues	Classic	Ctry	Elec	Folk	Jazz	Metal	Pop	Punk	Rock	Rap	Regg	Soul	a_i
Blues	0.68	0.10	**1.33**	0.11	0.28	0.57	0.17	0.66	0.15	0.92	0.09	0.04	0.06	*5.18*
Classical	0.07	0.03	0.18	0.03	0.04	0.06	0.15	0.25	0.10	**0.39**	0.01	0.01	0.01	*1.32*
Country	1.70	0.26	**6.03**	0.27	0.89	1.05	0.49	2.35	0.47	2.38	0.30	0.12	0.25	*16.56*
Electronic	0.11	0.04	0.28	0.12	0.08	0.10	0.27	0.48	0.24	**0.71**	0.05	0.05	0.02	*2.25*
Folk	0.20	0.04	**0.65**	0.07	0.23	0.16	0.07	0.27	0.08	0.42	0.02	0.02	0.02	*2.25*
Jazz	0.54	0.09	**0.90**	0.12	0.23	0.84	0.13	0.51	0.12	0.65	0.11	0.04	0.05	*4.32*
Metal	0.17	0.16	0.5	0.27	0.09	0.11	2.44	2.26	1.85	**4.06**	0.07	0.15	0.02	*12.16*
Pop	0.56	0.24	1.90	0.47	0.38	0.41	2.04	3.40	1.58	**5.41**	0.14	0.19	0.06	*16.77*
Punk	0.19	0.16	0.58	0.30	0.12	0.15	2.06	2.63	2.49	**4.02**	0.10	0.16	0.02	*12.98*
Rock	0.6	0.31	1.52	0.63	0.45	0.38	3.45	4.43	2.25	**7.06**	0.09	0.23	0.05	*21.46*
Reggae	0.16	0.04	0.41	0.06	0.05	0.18	0.10	0.37	0.12	0.43	**0.50**	0.06	0.06	*2.52*
Rap	0.03	0.02	0.10	0.05	0.02	0.03	0.15	0.24	0.11	**0.40**	0.06	0.10	0.01	*1.32*
Soul	0.05	0.01	**0.17**	0.01	0.03	0.05	0.02	0.08	0.02	0.14	0.02	0.01	0.01	*0.61*
b_j	5.05	1.49	14.54	2.52	2.90	4.10	11.55	17.93	9.57	27.00	1.55	1.18	0.63	100

Table 6.7 Normalised mixing matrix e^{CB} for the audio content-based network.

Network	Mixing coeff. r
CF	0.343
EX	0.411
CB	0.089

Table 6.8 Assortative mixing by genre coefficient r for the three networks, based on the matrices e in Tables 6.5, 6.6 and 6.7.

In the audio CB network, *country* and *rock* genres dominate over the rest. *Country* subsumes *blues*, *jazz* and *soul* genres. For instance, *folk* artists share a high fraction of links with *country* artists ($e^{CB}_{folk,country} = 0.65$, compared with $e^{CB}_{folk,folk} = 0.23$), yet $e^{CB}_{folk,rock}$ also presents a high correlation. This finding is aligned with our previous research presented in [8], where we conclude that *folk* and *country* genres are similar, using content-based audio similarity. Similarly, the same phenomenon happens for $e^{CB}_{blues,country}$, and $e^{CB}_{jazz,country}$, although in the latter case it is more arguably the similarity between the two genres.

Actually, in the CB network the bias towards *rock* and *country* genres is more prominent than in the two other networks. Artist similarity is derived from audio track similarity, thus preponderant genres have more chances to have links from other artists' genres. This is the reason why artists from infrequent genres correlate and "collapse" with the most prevalent ones (see Table 6.7).

Contrastingly, in the experts' network, *country*, *jazz* and *soul* artists present a high intra-correlation value (a high fraction of vertices linking artists of the same genre, $e^{EX}_{i,i}$). For instance, $e^{EX}_{jazz,jazz} = 11.71$, and the sum of the row (last column), a^{EX}_{jazz}, is 15.17. So, given a *jazz* artist, 77% of his similar artists are also *jazz* musicians ($\frac{e^{EX}_{jazz,jazz}}{a^{EX}_{jazz}} = 0.77$). Similar values are found for *country* and *soul* artists. Neither in CF nor in CB networks we can find these high intra-correlation values (only for the *reggae* genre in the CF network, with a $\frac{c_{reggae,reggae}^{CF}}{d_{reggae}^{CF}} = 0.66$ value).

At this point, we conclude the analysis of the three similar artist networks. Now, the following section presents the main findings about the correlation between artist popularity and their prominence in the similarity network.

6.2.3 Popularity Analysis

We have outlined in the previous section the main topological differences among the three networks. We add now the popularity factor (measured with the total playcounts per artist), by combining artists' rank in the Long Tail with the results from the network analysis. Two experiments are performed. The former reports the relationships among popular and unknown artists. The latter experiment aims at analysing the correlation between artists' indegree in the network and their popularity.

6.2.3.1 Artist Similarity

Figure 6.4 depicts the correlation among an artist's total playcounts and the total playcounts of its similar artists. That is, given the total playcounts of an artist (x axis) it shows, in the vertical axis, the average playcounts of its similar artists. CF network has a clear correlation ($r_{CF} = 0.503$); the higher the playcounts of a given artist, the higher the avg. playcounts of its similar artists. The *AMG* human expert network presents a moderate correlation ($r_{EX} = 0.259$). Thus, in some cases artists are linked according to their popularity. CB network does not present correlation ($r_{CB} = 0.08$). In this case, artists are linked independently of their popularity.

	$a_i \rightarrow a_j$	Head(%)	Mid(%)	Tail(%)
	Head	45.32	54.68	0
CF	Mid	5.43	71.75	22.82
	Tail	0.24	17.16	82.60
	Head	5.82	60.92	33.26
Expert	Mid	3.45	61.63	34.92
	Tail	1.62	44.83	53.55
	Head	6.46	64.74	28.80
CB	Mid	4.16	59.60	36.24
	Tail	2.83	47.80	49.37

Table 6.9 Artist similarity and their location in the Long Tail. Given an artist, a_i, it shows (in %) the Long Tail location of its similar artists (results are averaged over all artists). Each row represents, also, the Markov chain transition matrix for CF, CB, and expert-based methods.

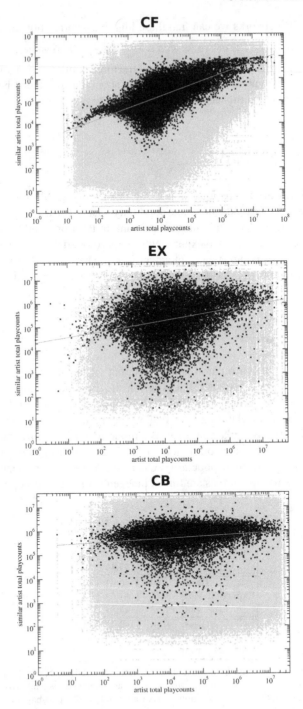

Fig. 6.4 A log–log plot depicting the correlation between an artist's total playcounts and similar artists' playcounts (average values are shown in *black*, *whilst grey dots* display all the values). Pearson correlation coefficient r values are: $r_{CF} = 0.503$, $r_{EX} = 0.259$ and $r_{CB} = 0.081$.

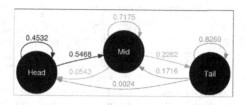

Fig. 6.5 Example of the Markov decision process to navigate along the Long Tail in the CF network. This information is directly derived from Table 6.9.

Table 6.9 presents artist similarity divided into the three sections of the Long Tail curve. Given an artist, a_i, it shows (in %) the Long Tail location of its similar artists (results are averaged over all artists). In the CF network, given a very popular artist, the probability of reaching (in one click) a similar artist in the tail is zero. Actually, half of the similar artists are located in the head part—that contains only 82 artists— and the rest are in the mid area. Artists in the mid part are tightly related (71.75%), and only 1/5 of the similar artists are in the tail part. Finally, given an artist in the tail, its similar artists remain in the same area. Contrastingly, the CB and EX networks promote the mid and tail parts much more in all the cases (specially in the head part).

Similarly to the mixing by genre, where we compute the correlation among the genres in linked artists, we can do the same for artist popularity. In fact, Table 6.9 directly provides us this information. For instance, given an artist in the *Head* part Table 6.9 shows the fraction of edges that are attached to the artist whose other ends are attached to artists of type *Head*, *Mid* or *Tail*. The mixing by popularity correlation coefficients are: $r_{CF} = 0.397$, $r_{EX} = -0.002$, and $r_{CB} = -0.032$. Again, the correlation values show that the CF network presents assortative mixing by popularity, whilst neither EX nor CB does.

6.2.3.2 From Head to Tail

To simulate a user surfing the recommendation network, we apply a Markovian stochastic process [10]. Indeed, each row in Table 6.9 can be seen as a Markov chain transition matrix, M, where the head, mid and tail parts are the different states. For example, Fig. 6.5 shows the Markov chain for the CF network. The values of matrix M denote the transition probabilities, $p_{i,j}$, between two states i, and j (e.g. $p_{head,mid}^{CF} = 0.5468$). The Markovian transition matrix, M^k, denotes the probability of going from any state to another state in k steps (clicks). The initial distribution vector, $P^{(0)}$, sets the probabilities of being at a determined state at the beginning of the process. Then, $P^{(k)} = P^{(0)} \times M^k$, denotes the probability distribution after k clicks, starting in the state defined by $P^{(0)}$.

Using $P^{(k)}$ and defining $P^{(0)} = (1_H, 0_M, 0_T)$, we can get the probability of reaching any state, starting in the head part. Table 6.10 shows the number of clicks needed to reach the tail from the head, with a probability $p_{head,tail} \geq 0.4$. In CF, one needs

	k	$\mathcal{P}^{(k)}$, with $P^{(0)} = (1_H, 0_M, 0_T)$	π	n
CF	5	$(0.075_H, 0.512_M, 0.413_T)$	$(0.044_H, 0.414_M, 0.542_T)$	26
Expert	2	$(0.030_H, 0.560_M, 0.410_T)$	$(0.027_H, 0.544_M, 0.429_T)$	8
CB	2	$(0.038_H, 0.562_M, 0.400_T)$	$(0.037_H, 0.550_M, 0.413_T)$	7

Table 6.10 Navigation along the Long Tail of artists in terms of a Markovian stochastic process. Second and third columns depict the number of clicks (k) to reach the tail from the head part, with a probability $p_{head,tail} \geq 0.4$. Fourth and fifth columns show the stationary distribution π, as well as the number of steps, n, to reach π (with an error $\leq 10^{-6}$).

five clicks to reach the tail, whereas in CB and expert-based only two clicks are needed.

Finally, the stationary distribution π is a fixed point (row) vector whose entries sum to 1, and that satisfies $\pi = \pi M$. The last two columns in Table 6.10 present the stationary distribution vector for each algorithm, and the number of steps to converge to π, with an error $\leq 10^{-6}$. CF transition matrix needs more than three times the number of steps of CB or EX to reach the steady state, due to the transition $p_{head,tail}^{CF} = 0$. Furthermore, even though the probability to stay in the tail in CF is higher than in CB or EX, this is due to the high probability to remain in the tail once it is reached ($p_{tail,tail}^{CF} = 0.8260$).

6.2.3.3 Artist Indegree

Up to now, we have analysed the popularity in terms of the relationships among the artists. Now, we analyse the correlation between the artists' indegree in the network and their popularity. As a starting point, we present in Table 6.11 the top-10 artists with the highest indegrees for each network. CF and expert-based contains two and eight mainstream artists, respectively. CF contains *U2* and *R.E.M.*, but the rest of the list contains more or less well known *jazz* musicians, including some in the top of the tail area. The whole list for the expert-based *AMG* network is made up of very popular artists. Our guess is that the editors connect long tail artists with the most popular ones, because these popular artists are considered influential and many bands are considered *followers of* these mainstream artists. The CB network has a more eclectic top-10 list, as one would expect. Oddly enough, there is no new or actual artists, but some classic bands and artists ranging several musical genres.Some bands are, in fact, quite representative of a genre (e.g. *Lynyrd Skynyrd*, and *The Charlie Daniels Band* for Southern-rock, *The Carpenters* for Pop in the 1970s, *George Strait* for Country, and *Cat Stevens* for Folk/Rock). Probably, their high indegree is due to being very influential in their respective musical styles. In some sense, there are other bands that "cite" or imitate their sound.

Although, the results could be somewhat biased; our sampled CF and expert networks are subsets of the whole *last.fm* and *AMG* similar artist networks, thus our sampling could not be a good representation of the whole dataset. Furthermore, the differences in the maximum indegree value (k_{in} for top-1 artist) among the three

CF		
k_{in}	Artist	Long Tail rank
976	Donald Byrd	6,362
791	Little Milton	19,190
772	Rufus Thomas	14,007
755	Mccoy Tyner	7,700
755	Joe Henderson	8,769
744	R.E.M.	88
738	Wayne Shorter	4,576
717	U2	35
712	Horace Silver	5,751
709	Freddie Hubbard	7,579

Expert		
k_{in}	Artist	Long Tail rank
180	R.E.M.	88
157	Radiohead	2
137	The Beatles	1
119	David Bowie	62
117	Nirvana	19
111	Tool	17
111	Pavement	245
109	Foo Fighters	45
104	Soundgarden	385
103	Weezer	51

CB		
k_{in}	Artist	Long Tail rank
1,955	George Strait	2,632
1,820	Neil Diamond	1,974
1,771	Chris Ledoux	13,803
1,646	The Carpenters	1,624
1,547	Cat Stevens	623
1,514	Peter Frampton	4,411
1,504	Steely Dan	1,073
1,495	Lynyrd Skynyrd	668
1,461	Toby Keith	2,153
1,451	Charlie Daniels Band	22,201

Table 6.11 Top-10 artists with higher indegree (k_{in}) for each recommendation network. The table shows too, the artist ranking in the Long Tail.

networks are due to the different sizes (N) and average degree $\langle k \rangle$ of the networks (5.47$_{EX}$ versus 14.13$_{CF}$, and 19.80$_{CB}$), but also due to the topology of the networks. CF and CB follow a power-law cumulative indegree distribution, whereas EX best fits a log-normal distribution. Therefore the maximum indegree k_{in} for EX is much smaller than that of CF or CB.

To conclude this analysis, Fig. 6.6 shows the correlation between artists' indegree (k_{in}), and artists' popularity, using artist's total playcounts. The figure shows whether the artists with higher indegree in the network (hubs) are the most popular artists. Again, we see that in CF and expert-based networks the artists with higher

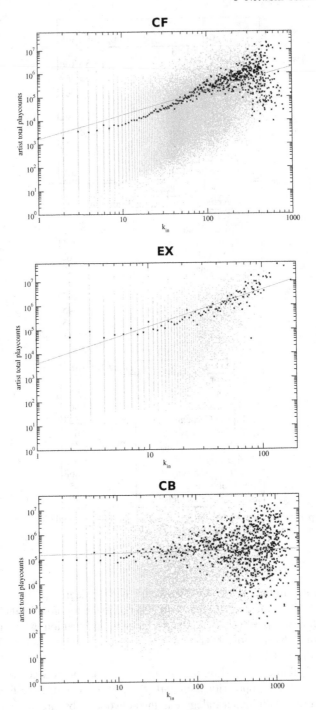

Fig. 6.6 A log–log plot showing the correlation between artist indegree (k_{in}, in horizontal axis) and its total playcounts (avg. values in *black*), in vertical axis. Pearson r values are: $r_{CF} = 0.621$, $r_{EX} = 0.475$, and $r_{CB} = 0.032$.

indegree (hubs) are mostly located in the head and mid part, whereas in CB they are more spread out through all the curve. Both CF and expert-based networks confirm the expectations, as there is a clear correlation between the artist indegree and total playcounts ($r_{CF} = 0.621$, and $r_{EX} = 0.475$). Artists with high indegree are the most popular ones. In CB, given a high indegree value it contains, on average, artists ranging different levels of popularity ($r_{CB} = 0.032$).

6.2.4 Discussion

The results show that the *last.fm* social-based recommender tends to reinforce popular artists, at the expense of discarding less-known music. Thus, the popularity effect derived from the community of users has consequences in the recommendation network. This reveals a somewhat poor discovery ratio when just browsing through the network of similar music artists. It is not easy to reach relevant long tail artists, starting from the head or mid parts (see Table 6.10). This could be related to the existence of positive feedback loops in social-based recommenders. The first users that enters to the system heavily affects the initial relationships among the items. After that, the users that come later, find an environment shaped by earlier users. These new users will be affected by the early raters that create the similarities among the items. Thus, positive feedback also affects the navigation through the Long Tail. Given a long tail artist, its similar artists are all located in the tail area as well. This does not always guarantee novel music recommendations; a user that knows an artist in the Long Tail quite well is likely to know most of the similar artists too (e.g. the solo project of the band's singer, collaborations with other musicians, and so on). Thus, these might not be considered good novel recommendations to that user, but familiar ones. CF contains, then, all the elements to conclude that popularity has a strong effect in the recommendations, because: (i) it presents assortative mixing (indegree–indegree correlation), see Fig. 6.3, (ii) there is a strong correlation between an artist's total playcounts and the total playcounts of its similar artists (see Fig. 6.4), (iii) most of the hubs in the network are popular artists (see Fig. 6.6), and (iv) it is not easy to reach relevant Long Tail artists, starting from the head or mid parts (see Table 6.10).

Human expert-based recommendations are more expensive to create and have a smaller Long Tail coverage compared to automatically generated recommendations like those in CF and CB. Regarding popularity, the hubs in the expert network are comprised of mainstream music, thus potentially creating a network dominated by popular artists (see Table 6.11 and Fig. 6.6). However, the topology—specially the log-normal cumulative indegree distribution—indicates that these artists do not act as hubs, as in the power law distributions with a γ exponent between 2 and 3 [7]. Furthermore, the expert network does not present assortative mixing (see Fig. 6.3), so artists are linked in a heterogeneous way; popular artists are connected with other less-known artists and the other way around (see Table 6.9 and Fig. 6.4).

According to the stationary distribution π (see Table 6.10), the key Long Tail area in the CB and EX networks are the artists in the mid part. These artists allow users to navigate inside the Long Tail acting as entry points, as well as main destinations when leaving the Long Tail. Also, users that listen to mainly very Long Tail music are likely to discover unknown artists—for them—that are in the mid part, and that are easily reachable from the artists in the tail. One should pay attention to the quality data in the Long Tail as well. Assuming that there exists some extremely poor quality music, CB is not able to clearly discriminate against it. In some sense, the popularity effect drastically filters all these low quality items. Although, it has been proved by [11] that increasing the strength of social influence increased both inequality and unpredictability of success and, as a consequence, popularity was only partly determined by quality.

6.3 User Network Analysis

One of the main goals of neighbourhood-based recommendation algorithms is to find like-minded people, and through them, discover unknown music. In this sense, a user similarity network resembles a social network, automatically connecting people that share similar interests.

We present an evaluation of two user similarity networks. Both networks are derived from the users' listening habits. The first one is based on collaborative filtering (CF). Again, we gather this information from *last.fm*. For the second network we use content-based audio similarity (CB) to compute user similarity.

6.3.1 Datasets

6.3.1.1 Social-Based, Collaborative Filtering Network

User similarity is gathered from *last.fm.*, using Audioscrobbler webservices. For each user we collect the top-20 similar users. *Last.fm* derives user similarity from the item-based approach, so it connects users that share common musical tastes. Table 6.12 shows the number of users and links in the network.

6.3.1.2 Content-Based Network

User similarity for the CB network is computed using content-based audio analysis from a music collection $(\mathcal{T})$ of 1.3 Million tracks of 30 s samples. To compute similar users we used all the tracks, $\mathcal{T}_u$, that a user u has listened to.

Distance between tracks, $d(x,y)$, is based on the Euclidean distance over a reduced space using Principal Component Analysis (PCA). The audio features used include not only timbral features (e.g. Mel frequency cepstral coefficients), but musical descriptors related to rhythm (e.g. beats per minute, perceptual speed, binary/ternary metric), and tonality (e.g chroma features, key and mode), among others [1]. Preliminary steps to compute the Euclidean distance are: (i) audio descriptor normalisation in the $[0,1]$ interval, and (ii) applying PCA to reduce the audio descriptors space to 25 dimensions. For each track, $t_i \in \mathcal{T}_u$, we obtain the most similar tracks:

$$sim(t_i) = \underset{\forall t \in \mathcal{T}}{\mathrm{argmin}}\left(d(t_i, t)\right), \tag{6.3}$$

Then, we get all the users, $\mathcal{U}_{sim(t_i)}$, that listened to any track similar to t_i. The list of (top-20) similar users of u is composed by the users in $\mathcal{U}_{sim(t_i)}$ for all $t_i \in \mathcal{T}_u$, weighted by the audio similarity distance:

$$similar\_users(u) = \bigcup \mathcal{U}_{sim(t_i)}, \forall t_i \in \mathcal{T}_u \tag{6.4}$$

To select the maximum number of similar users per user we compute, for all the users, the average distance between the user and her top-20 similar users. We use this average distance as a threshold to get the *top-N* most similar users, setting a maximum of $N = 20$.

The main difference between the two approaches is that in CF two users have to share at least one artist in order to become—potentially—similar. In the CB we can have two similar users that do not have share any artist, yet the music they listen to is similar. For instance, two users that listen to, respectively, $u_i = [Ramones, The Clash, Buzzcocks,$ and $Dead Kennedys]$, and $u_j = [Sex Pistols, The Damned, The Addicts,$ and $Social Distortion]$ could be very similar using CB similarity, but not using CF (unless the system also makes use of higher-level information, such as artist similarity derived from social tagging data).

However, due to the *similar_users(u)* equation we choose for the CB network, a user with a high number of songs in her profile has a higher chance of being considered similar to other users.

	Number of users	Number of relations
Last.fm social filtering (CF)	158,209	3,164,180
Content-based (CB)	207,863	4,137,500

Table 6.12 Datasets for the user similarity networks.

Property	CF (*last.fm*)	CB
N	158,209	207,863
$\langle k \rangle$	20	19.90
SGC	100%	99.97%
γ_{in}	NA (log-normal)	NA (log-normal)
$\langle d_{dir} \rangle (\langle d_{rand} \rangle)$	9.72 (3.97)	7.36 (4.09)
D	12	10
r	0.86	0.17
$C (C_{rand})$	0.071 (1.2^{-4})	0.164 (9.57^{-5})
$C(k) \sim k^{-\alpha}$	0.57	0.87

Table 6.13 User network properties for the *last.fm* collaborative filtering network (CF), and content-based audio filtering (CB). N is the number of nodes, and $\langle k \rangle$ the mean degree, $\langle d_d \rangle$ is the avg. shortest directed path, and $\langle d_r \rangle$ the equivalent for a random network of size N, D is the diameter of the (undirected) network. SGC is the size (percentage of nodes) of the strong giant component for the undirected network, γ_{in} is the power-law exponent of the cumulative indegree distribution (if applicable), r is the indegree–indegree Pearson correlation coefficient (assortative mixing), C is the clustering coefficient for the undirected network, C_r for the equivalent random network, and $C(k) \sim k^{-\alpha}$ is the α exponent for the clustering coefficient as a function of node degree (*scaling law*).

6.3.2 Network Analysis

6.3.2.1 Small World Navigation

Table 6.13 presents the properties of the two networks. The two networks moderately present the *small-world* phenomena [2]. They have a small average directed shortest path, $\langle d_d \rangle$, but higher than the $\langle d_r \rangle$ in the equivalent random network (twice as much). Also the two clustering coefficients, C, are significantly higher than the equivalent random networks C_r.

6.3.2.2 Clustering Coefficient

Figure 6.7 shows the clustering coefficient as a function of node degree $C(k)$, for the undirected network. We can see that the higher the indegree of a user, the lower her clustering coefficient. In this sense, the CB network resembles a hierarchical network [12], although it is not a *scale free* network. In a hierarchical network there are many small densely linked clusters that are combined to form larger but less cohesive groups, that a few prominent nodes interconnect. In our CB network, $C_{CB}(k) \sim k^{-0.87}$, starting at $k = 20$ the $\alpha = 0.87$ is close to the scaling law, $C(k) \sim k^{-1}$. The scaling law is used to determine the presence of hierarchy in real networks [12].

$C(k)$ is computed for the undirected networks. That is the reason that the $C_{CB}(k) \sim k^{-0.87}$ power law starts at $k = 20$. In the undirected network most of the nodes have $k \geq 20$—the node outlinks, k_{out}, plus the incoming links they receive

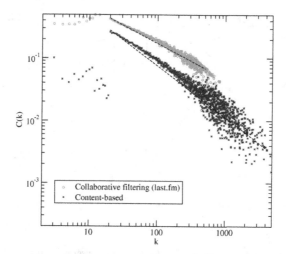

Fig. 6.7 Clustering coefficient $C(k)$ versus degree k. The CB network resembles a hierarchical network ($C_{CB}(k) \sim k^{-0.87}$), although it is not a *scale free* network.

k_{in}. However, in some cases a node has $k_{out} < 20$, because the threshold has been applied (see the creation of datasets, in Sec. 6.3.1). These few nodes are located on the left side of Fig. 6.7 ($0 < k < 20$), and are discarded to compute $C(k)$.

6.3.2.3 Indegree Distribution

Table 6.14 presents the model selection for the indegree distribution. For each network we give a *p-value* of the fit to the power-law model (first column). A higher *p-value* means that the distribution is likely to follow a power-law. We also present the likelihood ratios for the alternative distributions (power-law with an exponential cut-off, and log-normal), and the *p-values* for the significance of the likelihood ratio tests. In this case, a *p-value* close to zero means that the alternative distribution can also fit the distribution (see Sec. 4.4 for an in-depth explanation about fitting a probability density distribution, and the model selection procedure).

	power-law	power-law + cut-off		log-normal		support for
	p	LLR	p	LLR	p	power-law
CF	0.00	−192.20	0.00	−14.41	0.00	none
CB	0.00	−836.89	0.00	−37.05	0.00	none

Table 6.14 Model selection for the indegree distribution of the two user networks. For each network we give a *p-value* for the fit to the power-law model (first column). We also present the likelihood ratios for the alternative distributions (power-law with an exponential cut-off, and log-normal), and the *p-values* for the significance of each of the likelihood ratio tests (LLR).

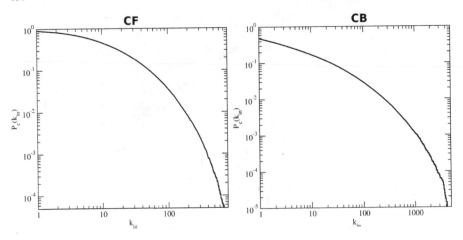

Fig. 6.8 Cumulative indegree distribution for the CF and CB user networks.

Figure 6.8 shows the cumulative indegree distribution for each network. Neither of the two networks are *scale free*, because the cumulative indegree distribution does not follow a power law (see Table 6.14, first column). In both networks the best fitting distribution, according to their log-likelihood, is a log-normal distribution. The best fit for the CF network is obtained with a log-normal distribution,

$f(x) = \frac{1}{x} e^{-\frac{(\ln(x)-\mu)^2}{2\sigma^2}}$. The parameters are mean of log $\mu = 6.49$, and standard deviation of log, $\sigma = 2.80$. The best fit for the CB network is obtained with a log-normal distribution. The parameters are mean of log $\mu = 8.51$, and standard deviation of log, $\sigma = 2.74$.

6.3.2.4 Assortative Mixing

Figure 6.9 depicts the assortative mixing—indegree indegree correlation—in the two user networks. CF presents assortative mixing, whilst CB does not ($r_{CF} = 0.86$ and $r_{CB} = 0.17$). The CF user similarity network resembles a social network, where it is very common the find *homophily*. Users with a high indegree, k_{in}, are connected to other users also with a high k_{in}, whereas users with a low indegree are connected to peers that also have a low indegree.

At this point, we conclude the analysis of the two user networks. The following section presents the analysis about the correlation between the user's location in the Long Tail of artist popularity and the user's prominence in the similarity network.

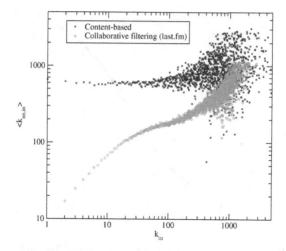

Fig. 6.9 Assortative mixing in the two user networks. CF presents assortative mixing, whilst CB does not ($r_{CF} = 0.86$ and $r_{CB} = 0.17$).

6.3.3 Popularity Analysis

Similar to the analysis performed in the artist networks, we present two experiments about the popularity effect in the user networks. The first reports the relationships among the users and their location in the Long Tail. The user's location in the Long Tail is measured by averaging the Long Tail location of the artists in the user profile. The second experiment analyses the correlation between users' indegree in the network and their location in the Long Tail.

6.3.3.1 User Similarity

To compute a user's location in the music Long Tail, we get the artists that user u listens to the most ($\mathcal{A}_u$). Summing all the artists' playcounts in $\mathcal{A}_u$ must hold at least 66% of the user's total playcounts, so it is a sound representation of the musical tastes of u. Then, the user's Long Tail location is computed as the weighted average of $\mathcal{A}_u$. That is, for each $a \in \mathcal{A}_u$ we combine the user playcounts for artist a with the Long Tail location of a. Figure 6.10 shows an example of a user's location in the Long Tail.

Interestingly, most of the users are located in the *Mid* part of the curve. Thus, on average a user listens to mainstream music (from the head and mid areas), but also some unknown bands. Because the *Mid* area is very dense, we split this part into three subsections: Mid_{top}, Mid_{middle}, Mid_{end}. Table 6.15 presents the user similarity in terms of Long Tail locations. The main difference between the two similarity networks is for the users in the *Head* part. In the CF network more than 55% of the

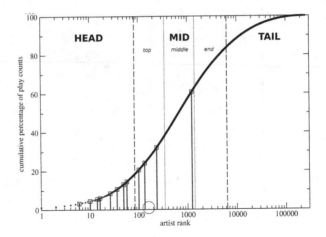

Fig. 6.10 Example of a user's location in the Long Tail of artists. The circle denotes the user's location, computed as the weighted average of the user profile artists' playcounts and popularity.

similar users are also located in the head part or in the top of the *Mid* part (*Mid$_{top}$*), whilst in the CB network this value is less than 35%.

	$u_i \rightarrow u_j$	**Head**	**Mid$_{top}$**	**Mid$_{middle}$**	**Mid$_{end}$**	**Tail**
	Head	9.36%	46.22%	26.66%	14.97%	2.78%
CF	**Mid**	1.11%	20.52%	41.96%	30.22%	6.18%
	Tail	0.41%	7.23%	26.98%	42.43%	22.95%
	Head	10.64%	23.70%	34.42%	25.91%	5.32%
CB	**Mid**	3.79%	15.43%	37.95%	34.92%	7.90%
	Tail	1.92%	8.34%	26.94%	40.81%	21.98%

Table 6.15 Similarities among the users, and their location in the Long Tail. Given a user, u_i, it shows (in %) the Long-Tail location of its similar artists, u_j. The results are averaged over all users in each part of the curve.

We represent each row in Table 6.15 as a Markov transition matrix. Using a Markovian stochastic process we can simulate a user surfing the similarity network. In the artist network (see Sec. 6.2.2), we were interested in the navigation from head to tail artists. Now, in the user network, the users are already located in the Long Tail according to the artists' popularity in their profile. Thus, we are more interested in the Long Tail location of the similar users, rather than in the navigation from head to tail users. For instance, using $P^{(3)}$ and defining $P^{(0)} = (0_{Head}, 0_{M\text{-}top}, 1_{M\text{-}mid}, 0_{M\text{-}end}, 0_{Tail})$, we get the probability of a user located in the mid part of the curve (*Mid$_{middle}$*) to move to the left side (*Head*, and M_{top}), to stay in the same *Mid$_{middle}$* area, or to move to the right (*Mid$_{end}$*, and *Tail*). Table 6.16 shows the probability distributions. Second column shows the probability distribution of a user located in the *Mid$_{middle}$* after 3 clicks, $P^{(3)}$. The CF network has a tendency to stay in the same Mid_{middle} area, whilst in the CB network the user slightly moves towards the right, tail, area. In both cases, the probability to move to the *Head* (left) is around 0.2.

	$\mathcal{P}^{(3)}$, with $P^{(0)} = (0,0,1,0,0)$	π	n
CF	$(0.210_{Left}, 0.407_{Stay}, 0.383_{Right})$	$(0.012_{Head}, 0.199_{M\text{-}top}, 0.407_{M\text{-}mid}, 0.309_{M\text{-}end}, 0.074_{Tail})$	5
CB	$(0.190_{Left}, 0.368_{Stay}, 0.442_{Right})$	$(0.039_{Head}, 0.151_{M\text{-}top}, 0.368_{M\text{-}mid}, 0.351_{M\text{-}end}, 0.091_{Tail})$	5

Table 6.16 Long Tail navigation in terms of a Markovian stochastic process. Second column shows the probability distribution of a user in the Mid_{middle} after 3 clicks. Third and fourth columns show the stationary distribution π, as well as the number of steps, n, to reach π (with an error $\leq 10^{-5}$).

Table 6.16 also shows the stationary distribution π, that satisfies $\pi = \pi M$. The last two columns present the stationary distribution vector for each algorithm, and the number of steps to converge to π, with an error $\leq 10^{-5}$. Both networks need the same number of steps to reach the steady state, confirming that overall the probability distributions are not very dissimilar.

6.3.3.2 User Indegree

We analyse the correlation between the users' indegree and their location in the Long Tail. Table 6.17 shows, for each network, the top-5 users with the highest indegrees. Users in the network with a high indegree can be considered as influential users or simply influentials. There is a big difference in the two networks; the influentials in CB are the users with the most playcounts, while the influentials in CF are the users that are closer to the *Head* part of the curve, independently of their total playcounts. In fact, only the top-4 users in the CF network have the same order of magnitude of total plays as the top-5 users in the CB network. Yet, around 60% of the CF top-4 user's playcounts correspond to *The Beatles*, the top-1 artist in the Long Tail of artist popularity. Therefore, the reason that CF top-4 user has a high indegree is not due to the high number of playcounts, but because most of the music she listens to is very mainstream.

Indeed, looking at the whole distribution of users—not only the top–5—in Fig. 6.11, the CF presents no correlation between the user's Long Tail position and their network indegree ($r_{CF} = -0.012$). However, CB network presents a correlation of $r_{CB} = 0.446$. Thus, as previously stated, users with higher indegree are the ones with the higher total playcounts in the CB network.

	k_{in}	LT Rank	Plays	Artists (number of plays)
	2,877	123	1,307	Arcade Fire (47), The Shins (43), Sufjan Stevens (42)
	2,675	75	2,995	Interpol (116), Arcade Fire (108), Radiohead (107)
CF	2,266	191	4,585	Broken Social Scene (172), Decemberists (128), Arch. Helsinki (128)
	2,225	176	38,614	The Beatles (23,090), The Doors (1,822), Bob Dylan (1,588)
	2,173	101	3,488	Decemberists (106), TV on the Radio (101), Arcade Fire (100)
	5,568	217	88,689	Red Hot Chili Peppers (27,618), Led Zeppelin (6,595), GN'R (3,457)
	4,706	789	105,768	Interpol (31,281), AFI (5,358), The Faint (3,056)
CB	4,207	1,222	21,762	Green Day (8,271), The Killers (4,040), The Strokes (2,184)
	3,991	121	77,433	The Cure (13,945), NIN (12,938), Smashing Pumpkins (8,460)
	3,884	550	44,006	Muse (19,178), The Killers (3,255), Green Day (3,168)

Table 6.17 Top-5 indegree (k_{in}) users. Influential users in CF are those located in the head of the Long Tail (column **LT Rank**), whilst influentials in CB are the ones with most playcounts (column **Plays**).

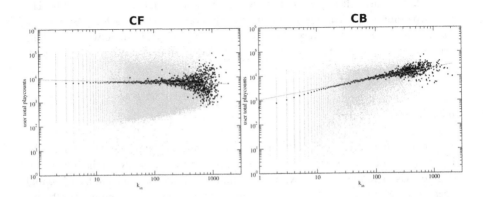

Fig. 6.11 Correlation between users' indegree and total playcounts. CB has a correlation of $r_{CB} = 0.446$, whilst CF does not present any correlation ($r_{CF} = -0.012$).

6.3.4 Discussion

The results of the analysis shows that the CB user similarity network resembles a hierarchical network (with the exception that CB is not a *scale-free* network). Thus, in the CB network there are a few nodes that are connecting smaller clusters. These nodes are the ones with the highest indegree which, according to Fig. 6.11, are the ones with higher total playcounts. Therefore, the users that listen to more music are the authorities in the CB network, independently of the quality or popularity of the music they listen to. This affects the navigation of the user similarity network. Contrastingly, in the CF network the users with a higher indegree are the ones that listen to more mainstream music. These users could have an impact for a recommender algorithm that uses user-based, instead of item-based, recommendations.

The key Long Tail area in the two user similarity networks is the *Mid* part. This area concentrates most of the users. To improve music discovery through user similarity, the recommendation algorithm should also promote users in the tail area. When computing user similarity, a recommender should take into account the users' location in the Long Tail curve.

An important missing aspect in our analysis is the dynamics of the user networks. It would be interesting to detect who are the tastemakers (or trendsetters). Users that create trends and have an impact in the musical tastes of other users are very relevant. This is related with the taxonomy of users presented in Sec. 3.2.1. Ideally, the *Savants* should be correlated with the tastemakers and influentials in the network. Detecting and tracking these users is a key point to improve music discovery through the network of similar users. However, detecting tastemakers can only be achieved by constantly gathering information about the users' music consumption. This way, we could analyse the dynamics and evolution of the user similarity network.

6.4 Summary

Recommender systems should assist us in the process of filtering and discovering relevant information hidden in the Long Tail. Popularity is the element that defines the characteristic shape of the Long Tail. We measure popularity in terms of total playcounts, and the Long Tail model is used in order to rank all music artists. We have analysed the topology and the popularity bias in two music recommendation scenarios; artist and user similarity. As expected by its inherent social component, the collaborative filtering approach is prone to popularity bias. This has some consequences on the discovery ratio, as well as navigation through the Long Tail.

Music recommender systems have to deal with biased datasets; a bias towards mainstream popular artists, towards a few prominent musical genres, or towards a particular type of user. Assortative mixing measures the correlation of these elements in the similarity network. In this sense, it is important to understand which contextual attributes have an impact when computing artist similarity (e.g. popularity, genre, decade, language, activity, etc.), or user similarity (e.g. age, race, language, etc.). The *Last.fm* social-based recommender presents several assortative mixing patterns. The artist network has assortative mixing on the nodes' indegree, but also presents mixing by genre, and mixing by popularity; i.e. the classical *homophily* issues that arise in social networks. Yet, as we will see in the next chapter, this does not necessarily have an impact on the quality of the recommendations.

The temporal effects in the Long Tail are another aspect one should take into account. Some new artists can be very popular, gathering a spike of attention when they release an album, but then they can slowly move towards the mid or tail area of the curve as time goes by. Thus, one-time hit items can be lost and forgotten in the Long Tail. Indeed, the music back-catalogue located in the Long Tail is an example of old and forgotten items that offer the possibility to be re-discovered by the users. A recommender system should be able to present and recommend these items to the user.

6.4.1 Links with the Following Chapters

We have presented a network-centric analysis of the similarities between artists, and between users. The network-based approach does not put the user into the evaluation loop. Without any user intervention it is impossible to evaluate the quality and user satisfaction of the recommendations, which does not necessarily correlate with predicted accuracy [13]. So, we still need to evaluate the quality of the recommendations as well as the popularity effect when providing recommendations to the users. For this reason, we present the user-based evaluation in the next chapter.

References

1. P. Cano, M. Koppenberger, and N. Wack, "An industrial-strength content-based music recommendation system," in *Proceedings of 28th International ACM SIGIR Conference*, (Salvador, Brazil), 2005.
2. D. J. Watts and S. H. Strogatz, "Collective dynamics of 'small-world' networks," *Nature*, vol. 393, pp. 440–442, June 1998.
3. J. M. Kleinberg, "Navigation in a small world," *Nature*, vol. 406, p. 845, 2000.
4. M. E. J. Newman, "The structure and function of complex networks," *SIAM Review*, vol. 45, no. 2, pp. 167–256, 2003.
5. A. Clauset, C. R. Shalizi and M. E. J. Newman, "Power-law distributions in empirical data," *SIAM Reviews*, June 2007.
6. A.-L. Barabási, R. Albert, H. Jeong, and G. Bianconi, "Power-law distribution of the world wide web," *Science*, vol. 287, p. 2115a, 2000.
7. A. L. Barabási and R. Albert, "Emergence of scaling in random networks," *Science*, vol. 286, pp. 509–512, October 1999.
8. M. Sordo, O. Celma, M. Blech, and E. Guaus, "The quest for musical genres: Do the experts and the wisdom of crowds agree?" in *Proceedings of the 9th International Conference on Music Information Retrieval*, (Philadelphia, PA), 2008.
9. K. Jacobson and M. Sandler, "Musically meaningful or just noise? an analysis of on-line artist networks," in *Proceedings of the 6th International Symposium on Computer Music Modeling and Retrieval*, (Copenhagen, Denmark), 2008.
10. S. P. Meyn and R. L. Tweedie, *Markov Chains and Stochastic Stability*. London: Springer, 1993.
11. M. J. Salganik, P. S. Dodds, and D. J. Watts, "Experimental study of inequality and unpredictability in an artificial cultural market," *Science*, vol. 311, pp. 854–856, February 2006.
12. E. Ravasz and A. L. Barabási, "Hierarchical organization in complex networks," *Physical Review. E, Statistical, Nonlinear, and Soft Matter Physics*, vol. 67, February 2003.
13. S. M. McNee, J. Riedl, and J. A. Konstan, "Being accurate is not enough: how accuracy metrics have hurt recommender systems," in *Computer Human Interaction. Human Factors in Computing Systems*, (New York, NY), pp. 1097–1101, ACM, 2006.

Chapter 7
User-Centric Evaluation

Up to now, we have presented a user agnostic network-based analysis of the recommendations. In this chapter we present a user-centric evaluation of the recommender algorithms. This user-based approach focuses on evaluating the user's perceived quality and usefulness of the recommendations. The evaluation method considers not only the subset of items that the user has interacted with, but also the items outside the user's profile. The recommender algorithm predicts recommendations to a particular user—taking into account her profile—and then the user provides feedback about the recommended items. Figure 7.1 depicts the approach.

7.1 Music Recommendation Survey

We aim at measuring the novelty and perceived quality of music recommendation, as neither system- nor network-centric approaches can measure these two aspects. However, we need to explicitly ask the users whether they already know the provided recommendations or not.

The proposed experiment is based on providing song recommendations to users, using three different music recommendation algorithms. Feedback gathered from the users consists of (i) whether a user already knows the song, and (ii) the relevance of the recommendations—whether she likes the recommended song or not.

7.1.1 Procedure

We designed a web-based survey experiment to evaluate the novelty and relevance of music recommendations from the point of view of the users. The survey is divided in two sections. The first one asks the participants for basic demographic

Ò. Celma, *Music Recommendation and Discovery*,
DOI 10.1007/978-3-642-13287-2_7, © Springer-Verlag Berlin Heidelberg 2010

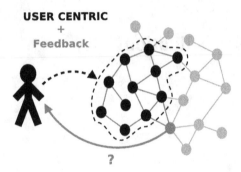

Fig. 7.1 User-centric evaluation focuses on evaluating the user's relevance and usefulness of the recommendations. The evaluation method considers not only the subset of items that the user has interacted with, but also the items outside the user's profile.

information (age range and gender), previous musical knowledge, and the average number of listening hours per day. The second part of the survey provides a set of rounds, each round containing an unsorted list of ten recommended songs evenly distributed from three different recommendation approaches. The participants do not know which recommendation method is used to recommend each song. A participant has to rate at least 10 songs, but she can rate as many songs as she likes.

The participant's feedback includes whether she knows the song (*no*, recall *only the artist*, recall *artist name and song title*), and the quality of the recommendations —whether she likes the song or not—on a rating scale from 1 (*I don't like it*) to 5 (*I like it very much*). The recommended songs do not contain any metadata, neither artist name nor song title, but only an audio preview of 30 s. The participant can listen to the preview of the recommended song as many times as she wishes. Figure 7.2 shows a screenshot of the experiment.

7.1.2 Datasets

The three music recommendation algorithms used are: collaborative filtering (CF), content-based audio similarity (CB), and a hybrid approach (HY) combining *Allmusic.com* human expert information, and content-based similarity. CF song similarity comes, again, from *last.fm*,[1] using the *Audioscrobbler* web services (API v1.0). The CB method is the one explained in Sec. 6.2.1, Eq. (6.1). Hybrid method (HY) is based on combining related artists from *Allmusic.com* musicologists, and CB audio similarity at track level. That is, to get the similar tracks from a seed track, first it gets the related artists (according to the *AMG* human experts) of the artist's seed

---

[1] See for example http://www.last.fm/music/U2/_/One/+similar

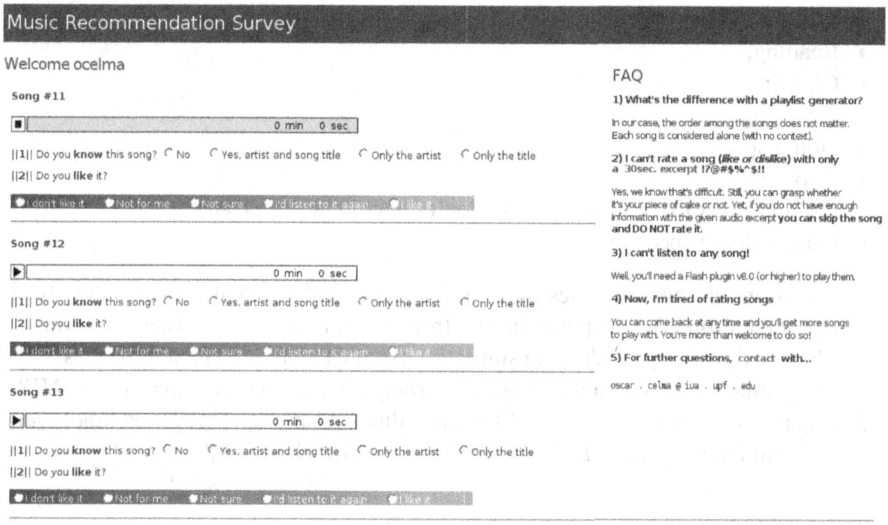

Fig. 7.2 Screenshot of the Music recommendation survey.

track. Then, it ranks the retrieved tracks from the related artists using content-based audio similarity with the seed track.

7.1.3 Participants

In order to characterise the participants, at the beginning of the survey they were asked to provide some basic demographic information (age range, and gender), as well as the participants musical background knowledge, the average number of listening hours per day (*more than 4 h a day, between 2 and 4 h a day, less than 2 h a day, almost never listen to music*), and the context while listening to music. All the fields were optional, so the participants could fill-in or not the information (only 9 participants did not fill-in all the data). Regarding the musical background, the survey offered the following single choice options:

- *None*: no particular interest in music related topics.
- *Basic*: lessons at school, reading music magazines, blogs, etc.
- *Advanced*: regular choir singing, amateur instrument playing, remixing or editing music with the computer, etc.
- *Professional*: professional musician—conductor, composer, high level instrument player—music conservatory student, audio engineer, etc.

Regarding the context while listening to music, the participants were asked to choose (multiple selection was allowed) the situations were they often listen to music. The options are:

- While working,
- Reading,
- Cleaning,
- Traveling,
- Doing sport,
- Cooking,
- Usually I just listen to music (and don't do anything else), and
- Other (please specify)

Furthermore, musical tastes of the participants were modelled using some seed tracks of their top-20 most played artists from their *last.fm* profile. These seed tracks are the ones used to provide song similarity using CF, CB and HY approaches.

To assemble a significant number of participants, we sent an email to the *MIR-list*[2] that described the survey and the procedure. Also, the survey was kindly announced in Paul Lamere's *Duke Listens* blog[3] on March 3rd, 2008.

7.2 Results

After running the experiment during the first two weeks in March 2008, 5,573 tracks were rated by 288 participants (with an average of 19 tracks rated per participant). Section 7.2.1 presents the analysis of the participants' data. Then, Sect. 7.2.2 presents the results of the three music recommendation approaches, including the analysis of the perceived quality, as well as the novelty and familiarity elements.

7.2.1 Demographic Data

We present the results of the demographic and musical background data gathered from the participants. Figure 7.3 shows the information about the participants' demographics. Most of the participants were adult males between 19 and 35 years old.

Figure 7.4 shows the distribution of the participants' musical background. Participants had a basic or advanced musical background, and most of them spent an average of two or more hours per day listening to music. The four pie charts have a 3% of not-available (*NA*), missing data. This missing data comes from nine participants that answered none of the questions.

To recap, our predominant participants were male young adults, with a basic or advanced musical background, who listen to quite a lot of music during the day. We consider that this is a biased sample of the population of listeners open to receiving music recommendations. Yet, it is the group we could engage to answer the survey.

---

[2] Message sent to music-ir@listes.ircam.fr on February, 28th, 2008

[3] http://blogs.sun.com/plamere/entry/evaluating_music_recommendations

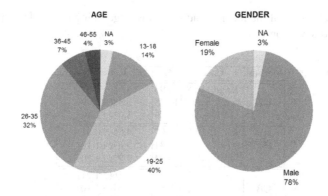

Fig. 7.3 Demographic information (age and gender distribution) of the participants.

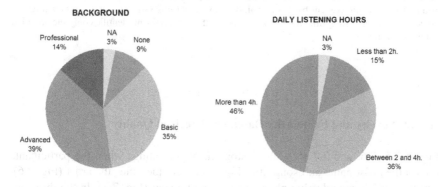

Fig. 7.4 Musical background and daily listening hours information of the participants.

7.2.2 Quality of the Recommendations

Now, we present the results of the second part of the survey, which consists on the evaluation of the three music recommendation methods. During the experiment, a list of 5,573 tracks rated by 288 participants was compiled. Feedback for each recommended song includes whether the user identifies the song (*no*, recall *only the artist*, recall *artist name and song title*), and the relevance of the recommendation (on a [1..5] scale) based on the 30 s audio excerpt.

7.2.2.1 Overall Results

Table 7.1 presents the overall results for the three algorithms. It shows, for each algorithm, the percentage of recommended songs that the participants identified (i.e. they are familiar with), as well as the unknown—novel—ones. The last column shows the relevance of the recommendations (average rating in a scale of [1..5], and standard deviation).

Method	Case	%	Avg. Rating (Stdev)
CF	*Recall A&S*	14.93	4.64(±0.67)
	Recall only A	12.23	3.88(±0.99)
	Unknown	71.69	3.03(±1.19)
HY	*Recall A&S*	10.07	4.55(±0.81)
	Recall only A	10.31	3.67(±1.18)
	Unknown	78.34	2.77(±1.20)
CB	*Recall A&S*	9.91	4.56(±1.21)
	Recall only A	7.95	3.61(±1.10)
	Unknown	80.97	2.57(±1.19)

Table 7.1 User-centric evaluation of the novelty component for collaborative filtering (CF), Hybrid (HY), and audio content-based (CB) algorithms. *Recall A&S* means that a participant recognises both artist and song title. *Recall only A* means that a participant identifies only the artist but not the song title.

7.2.2.2 Novelty and Familiarity Based on Perceived Quality

Figure 7.5, 7.6, and 7.7 show the histogram of the ratings when the participants knows the artist name and song title (Fig. 7.5), only identifies the artist (Fig. 7.6), and the song is completely unknown to the participant (Fig. 7.7). In the three approaches, familiar recommendations score very high; specially when the participant identifies the song, but also when it only recognises the artist. Yet, providing familiar recommendations is not the most challenging part of a recommender system. In fact, one can always play songs from the artists in the user's profile, but then the discovery ratio will be null.

As expected, the quality of the ratings drastically decrease when the participantis do not recognise the recommendations. The worst case is on the novel songs. Only the CF approach has an average rating score above 3 (see Table 7.1, and the box-and-whisker plots in Fig. 7.8). These bad results are comprehensible because in the experiment we intentionally did not provide any context about the recommendations, not even basic metadata such as the artist name or song title. One of the goals of the experiment is also to measure the novelty component, so the only input the participants can receive is the audio content. Our belief is that adding basic metadata and an explanation of why the song was recommended, the perceived relevance of the novel songs could be drastically increased in the three algorithms.

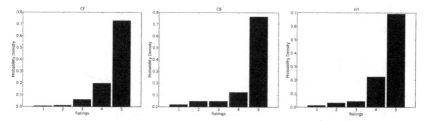

Fig. 7.5 Histogram of the ratings (on a [1..5] scale) when the participant identifies the artist and song (left: CF, center: CB, and Right: HY).

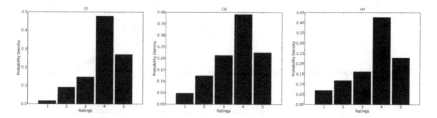

Fig. 7.6 Histogram of the ratings (on a [1..5] scale) when the participant only recognises the artist (left: CF, center: CB, and Right: HY).

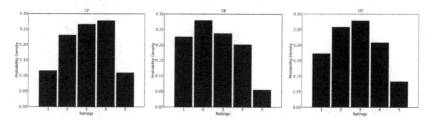

Fig. 7.7 Histogram of the ratings (on a [1..5] scale) when the recommended song is unknown to the participant (left: CF, center: CB, and Right: HY).

7.2.2.3 Analysis of Variance

We use the overall results from Table 7.1 to compare the three algorithms, performing a (non-parametric) Kruskal–Wallis one-way ANOVA within subjects, at 95% confidence level. As for familiar recommendations (including both *artist and song known* and recall *only artist*), there is no statistically significant difference in the relevance of the recommendations for the three algorithms. The main differences are found in the ratings of unknown songs, $F = 29.13$, with $p \ll 0.05$, and in the percentage of known songs, $F = 7.57$, $p \ll 0.05$. In the former case, the Tukey's test for pairwise comparisons confirms that CF average rating scores higher than HY and CB, at 95% family-wise confidence level (see Fig. 7.8 and 7.9). However, according to the latter case (percentage of known songs), CF generates more famil-

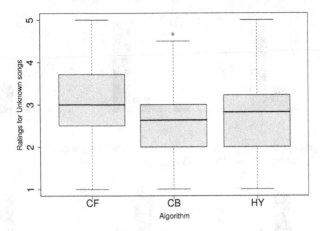

Fig. 7.8 Box-and-whisker plot for the ratings of unknown songs.

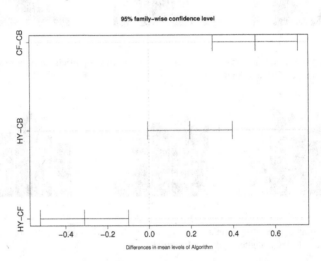

Fig. 7.9 Tukey's test for the ratings of unknown songs. Tukey's test does a pairwise comparison of the average ratings of unknown songs, and it confirms that CF avg. rating scores higher than HY and CB approaches, at 95% family-wise confidence level.

iar songs than CB and HY. Thus, CB and HY provide more novel recommendations, although their quality is not as good as CF.

7.3 Discussion

The results from the user-centric evaluation show that user perceived quality for novel, unknown recommendations —in the three methods—is on the negative side (avg. rating around 3/5 or less, in Table 7.1). This emphasises the need for adding more context when recommending unknown music. Users might want to understand why a song was recommended to them. Recommender systems should give as many reasons as possible, even including links to external sources (reviews, blog entries, etc.) to support their decision. Besides, the limitation in the experiment of using only 30 sec. samples did not help to assess the quality of the song. Yet, there are lots of industrial music recommender systems that can only preview songs due to licensing constraints. This constraint, then, is not that far from the reality.

We were expecting some correlation between the users musical background and the ratings or percentage of unknown songs. For instance, a user that listens to many hours of music daily could have more chances to identify more recommended songs. Yet, no big statistically significant differences were found, regarding the age, gender, musical background, number of hours, or context when listening to music. Only two minor statistically significant findings were found, with a p-value $p \ll 0.05$. The first one is that participants aging 36–45 (7% of the total) give lower ratings for the known songs than the rest of the participants. The second finding is that participants with no musical background (9% of the total) are the ones that penalise the unkonwn songs with lower ratings. Yet, these two results could have appeared by chance, given the low percentage of these two groups of participants.

An interesting experiment would be to identify each participant as a *savant, enthusiast, casual* or *indifferent* (see Sect 3.2.1), and see whether there is any difference in the ratings when providing novel music. This would measure how open to receiving novel recommenations each type of user is. Indeed, this would help music recommender systems to decide whether being risky or confident with the personalised recommendations. However, with the participants data that we gatehered it was not straightforward to decide which type of user each participant was.

Regarding recommendation approaches, the context-free and popularity agnostic CB algorithm sometimes points in the wrong direction (it is not that easy to discriminate between a, say, classical guitar and a harpsichord, based solely on the audio content), and gives poor or non-sense recommendations. This leaves room for improving the audio similarity algorithm. In this sense, the proposed hybrid approach drastically reduces the space of possible similar tracks to those artists related to the original artist. This avoids, most of the time, the *mistakes* performed by the pure CB, but on the other hand the HY results are less *eclectic* than CB. CF tends to be more conservative, providing less novel recommendations, but of higher quality, relevant to the user. Figure 7.10 summarises the comparison of the three approaches, based on the trade-off between novelty and relevance (presented in Chap. 4, Fig. 4.8).

We can envision different solutions to cope with novelty in recommender systems. The first one is to use CF, promoting unknown artists by means of exploiting the Long Tail popularity of the catalog and the topology of the recommendation

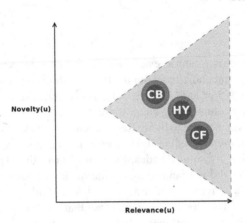

Fig. 7.10 Location of the three music recommendation approaches in the novelty vs. relevance axis (presented in Chap. 4, Fig. 4.8).

network. Another option is switching among algorithms when needed. For instance, to avoid the cold-start problem whilst promoting novelty, one option is to use CB or the hybrid approach, although this one heavily relies on human resources. After a while, the system can move to a stable CF or HY approaches. Or, we could also take into account the artist's (or user) location in the Long Tail, and use one or another algorithm accordingly. Furthermore, the system should be able to change the recommendation approach according to the user's needs. Sometimes, a user is open to discovering new artists and songs (novelty), while sometimes she just wants to listen to her favourites (familiarity). Detecting these modes and acting accordingly should increase the user's satisfaction with the system.

7.4 Limitations

To conclude, we also want to point out some limitations of the experiment. Users had to rate songs using only a 30 s audio preview. Even though the participants could listen to the songs repeatedly, it is not easy to rate a song the first time one listens to it. Sometimes, one can love a song after hearing it several times, in different contexts and moods. We could not measure this effect in the experiment. One solution could be to allow participants to download the full songs, and then after a period of time (e.g. 1 week, 1 month) they notify us with the total playcounts for each recommended song. Relevant songs could be inferred from the listening habits about the recommended songs. However, in this case a limitation is that we would collect less answers from the participants (i.e. only the songs that were listened to at least once).

The results are presented sequentially. Even though we clearly stated that the result is not playlist, thus there is no order. Still, the previously listened tracks can influence the user opinion of the current song. Also, if the first songs seem inappropriate, the songs displayed afterwards may seem better than they actually are.

Another issue is that musical tastes from the participants were gathered from *last.fm*, which is also one of the recommendation approaches used. This means that, beforehand, the participants were used to this system and the recommendations it provides. Yet, we decided that this music profile is more compact and reliable than asking the participant, at the beginning of the experiment, to enter a list of her favourite artists. Furthermore, another constraint is that only users with a *last.fm* account could participate in the survey.

The blind recommendation method approach—without providing any context—does not help in assessing the relevance of the novel recommendations. It might be the case that some of the novel songs were rated badly, but when explaining the relationships with the user's favourite artists, the artist biography, images, etc. the perceived quality could be increased. In real recommender systems, blind recommendations with no explanations are useless. *Why* is as important as *what* is being recommended.

Last but not least, we are not interested on judging which recommendation method performs the best, but on detecting the main differences among the approaches, and how people respond to each approach. In this sense, it is not fair to compare a real system like *last.fm* to the other two straight-forward plain approaches. In addition, we did not include a fourth method, say a random recommender, that could serve us as a baseline for the recommendations. This way, we could assess whether the three methods perform, at least, better than the baseline. Instead, we chose to gather more ratings from the three real methods than adding another—baseline—method in the survey.

Chapter 8
Applications

This chapter presents two implemented prototypes that are related with the main topics presented in the book; music discovery and recommendation. The first system, named, *Searchsounds*, is a music search engine based on text keyword searches, as well as a *more like this* button, that allows users to discover music by means of audio similarity. Thus, *Searchsounds* allows users to dig into the Long Tail, by providing music discovery using audio content-based similarity. The second system, named *FOAFing the Music*, is a music recommender system that focuses on the Long Tail of popularity, promoting unknown artists. The system also provides related information about the recommended artists, using information available on the web gathered from music related RSS feeds.

The main difference between the two prototypes is that *Searchsounds* is a non-personalised music search engine, whilst *FOAFing the Music* takes into account the user profile and the listening habits to provide personalised recommendations.

8.1 Searchsounds: Music Discovery in the Long Tail

Searchsounds, is a web-based music search engine that allows users to discover music using content-based similarity. Section 8.1.1 introduces the motivations and background of the system implemented. In Sec. 8.1.3 we present the architecture of the system. Finally, the last section summaries the work done and outlines the remaining work regarding the functionality of the system.

8.1.1 Motivation

Nowadays, the increasing amount of available music in the World Wide Web makes very difficult, to the user, to find music she would like to listen to. To overcome this problem, there are some audio search engines that can fit the user's needs.

Ò. Celma, *Music Recommendation and Discovery*,
DOI 10.1007/978-3-642-13287-2_8, © Springer-Verlag Berlin Heidelberg 2010

Some of the current existing search engines are, nevertheless, not fully exploited because their companies would have to deal with copyright infringing material. As general search engines, music search engines have a crucial component: an audio crawler, that scans the web for audio files, and also gathers related information about files [1].

8.1.1.1 Syndication of Web Content

During the last years, syndication of web content—a section of a website made available for other sites to use—has become a common practice for websites. This originated with news and weblog sites, but nowadays is increasingly used to syndicate any kind of information. Since the beginning of 2003, a special type of weblog, named audio weblogs (or MP3 blogs), has become very popular. These blogs make music titles available for download. The posted music is explained by the blog author, and usually it has links that allow users to buy the complete album or work. Sometimes, the music is hard to find or has not been issued in many years, and many MP3 blogs link strictly to music that is authorised for free distribution. In other cases, MP3 blogs include a disclaimer stating that they are willing to remove music if the copyright owner objects. Anyway, this source of semi-structured information is a jewel for web crawlers, as it contains the user's object of desire—the music—and some textual information that is referring to the audio file.

The file format used to syndicate web content is XML. Web syndication is based on the RSS family and Atom formats. The RSS abbreviation is used to refer to the following standards: Really Simple Syndication (RSS 2.0), Rich Site Summary (RSS 0.91 and 1.0) or RDF Site Summary (1.0).

Of special interest are the feeds that syndicate multimedia content. These feeds publish audiovisual information that is available on the net. An interesting example is the Media RSS (mRSS) specification,[1] lead by *Yahoo!* and the multimedia RSS community. mRSS allows bloggers to syndicating multimedia files (audio, video, image) in RSS feeds, and adds several enhancements to RSS enclosures. Although mRSS is not yet widely used on the net, some websites syndicate their multimedia content following the specification. These feeds contain textual information, plus a link to the actual audiovisual file. As an example, Listing 8.1 shows a partial RSS feed.[2]

```
<rss version="2.0"
xml:base="http://www.ourmedia.org"
xmlns:media="http://search.yahoo.com/mrss"
xmlns:dc="http://purl.org/dc/elements/1.1/"
>
<channel>
 <title>Example of a mRSS feed</title>
 <link>http://www.ourmedia.org/user/45801</link>
```

---

[1] http://search.yahoo.com/mrss/

[2] Adapted from a real example published in *OurMedia* website. http://www.ourmedia.org

```
<description>
 Recently published media items from Ourmedia.org
</description>
<language>en</language>
<item>
 <title>Fanky beats</title>
 <link>http://www.ourmedia.org/node/...</link>
 <description>Rock music with a funky beat and electric lead
     guitar riffs (...)</description>
 <pubDate>Mon, 17 Apr 2007 01:35:49 -0500</pubDate>
 <dc:creator>John Brettbutter</dc:creator>
 <category domain="urn:ourmedia:term:35">
  Alternative Rock
 </category>
 <category domain="urn:ourmedia:term:582">funk</category>
 <category domain="urn:ourmedia:term:727">guitar</category>
 <enclosure url="http://archive.org/.../file.mp3"
  length="3234212" type="application/octet-stream" />
 </item>
 <item>
 <title>Another item</title>
 ...
 </item>
</channel>
</rss>
```

Listing 8.1 Example of a media RSS feed.

The example shows an item with all its information: the title of the item, the description, the publication date, the editor of the entry, and a set of categories (similar to tags, but controlled from a given taxonomy). *Searchsounds* mines this information in order to retrieve relevant audio files based on keywords.

8.1.2 Goals

The main goal of the system is to allow users to discover unknown music. For this reason, *Searchsounds* mines music related information available in MP3-weblogs, and attaches textual information to the audio files. This way, users can search and retrieve music related to the query, as well as music that sounds similar to the retrieved audio files. This exploration mode allows users to discover music—related to his original (keyword based) query—that would be more difficult to discover using only textual queries.

Figure 8.1 shows the relationship between the music information plane (see Sec. 3.3), and the information that *Searchsounds* uses.

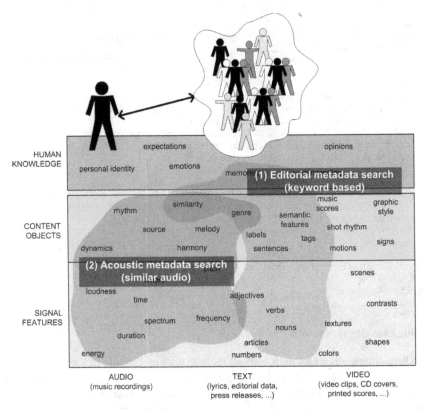

Fig. 8.1 *Searchsounds* makes use of editorial, cultural and acoustic metadata. The system retrieves (1) audio files from a keyword query, as well as (2) a list of (content-based) similar titles.

8.1.3 System Overview

Searchsounds exploits and mines all the music related information available from MP3-weblogs. The system gathers editorial, cultural, and acoustic information from the crawled audio files. The input of the system is a query composed by text keywords. From these keywords, the system is able to retrieve a list of audio files related with the query. Each audio file provides a link to the original weblog, and a list of similar titles. This similarity is computed using content-based audio description. Thus, from the results of a keyword query, a user can discover related music by navigating onto the audio similarity plane. It is worth to mention that there is no user profiling or any kind of user representation stored in the system. This is a limitation, as the system does not make any personalised recommendations. However, this limitation is solved in the next prototype (explained in Sec. 8.2). The main components of the system are the audio crawler and the audio retrieval system. Figure 8.2 depicts the architecture of the system.

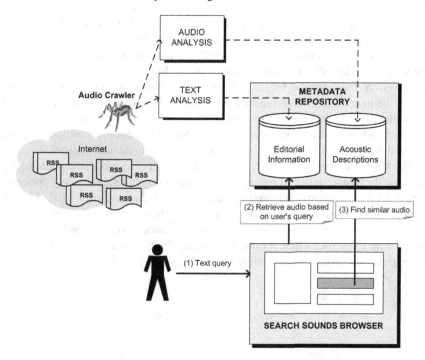

Fig. 8.2 *Searchsounds* architecture. The main components are the audio crawler, and the audio retrieval system.

8.1.3.1 Audio Crawler

The system has an audio spider module that crawls the web. All the gathered information is stored into a relational database. The audio crawler starts the process from a manually selected list of RSS links (that point to MP3-blogs). Each RSS file contains a list of entries (or *items*) that link to audio files. The crawler seeks for new incoming items—using the *pubDate* item value and comparing with the latest entry in the database—and stores the new information into the database. Thus, the audio crawler system has an historic information of all the items that appeared in a feed.

From the previous RSS example (see Example 8.1, presented in Section 8.1.1.1), the audio crawler stores the *title*, the content of the *description*, the assigned terms from the taxonomy (*category* tags), and the link to the audio file (extracted from the *enclosure url* attribute).

8.1.3.2 Audio Retrieval System

The logical view of a crawled feed item can be described by the bag-of-words approach: a document is represented as a number of unique words, with a weight (in our case, the tf/idf function) assigned to each word [2]. Special weights are assigned to the music related terms, as well as the metadata (e.g. ID3 tags) extracted from the audio file. Similar to our approach, [3] presents a proposal of modifying the weights of the terms pertaining to the musical domain.

Moreover, basic natural language processing methods are applied to reduce the size of the item description (elimination of stopwords, and apply Porter's stemming algorithm [4]). The information retrieval (IR) model used is the classic vector model approach, where a given document is represented as a vector in a multidimensional space of words (each word of the vocabulary is a coordinate in the space).

The similarity function, $sim(d_j, q)$, between a query (q) and a document (d_j) is based on the cosine similarity, using $TF\text{-}IDF$ weighting function (already presented in Sec. 2.5.4). Our approach is well suited not only for querying via artists' or songs' names, but for more complex keyword queries such as: *funky guitar riffs* or *traditional Irish tunes*. The retrieval system outputs the documents (i.e. feed entries) that are relevant to the user's query, ranked by the similarity function. Figure 8.3 depicts the retrieved audio files for *traditional Irish music* query.

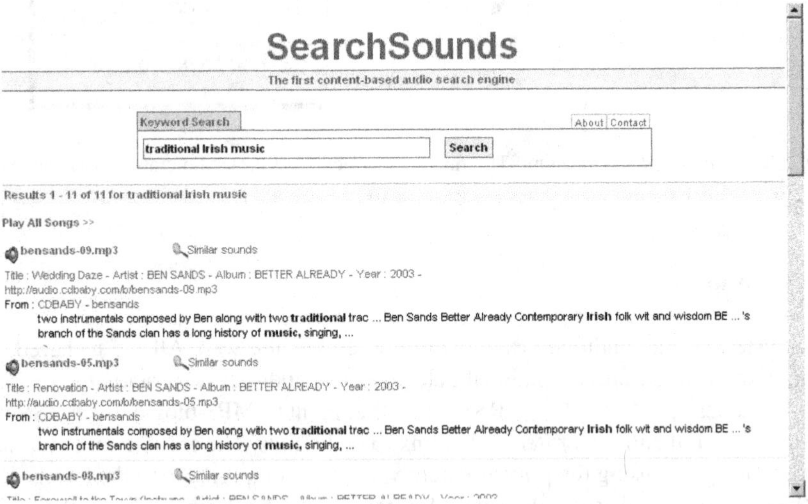

Fig. 8.3 Screenshot of the *Searchsounds* application, showing the first 10 results from *traditional Irish music* query.

Based on the results obtained from the user's textual query, the system allows users to find similar titles using content-based audio similarity. Each link to an audio file has a *Find similar* button that retrieves the most similar audio files, based on a set of low and mid-level audio descriptors. These descriptors are extracted from the audio and represent properties such as: rhythm, harmony, timbre and instrumentation, intensity, structure and complexity [5].

This exploration via browsing allows users to discover music—related to his original (keyword based) query—that would be more difficult to discover by using textual queries only. There is an analogy between this type of navigation and, for example, Google's "find web pages that are similar to a given HTML page". In our case, similarity among items are based on audio similarity, whereas Google approach is based on the textual content of the HTML page. Still, both browsing approaches are based on the *content* analysis of the retrieved object.

8.1.4 Summary

We developed a web-based audio crawler that focuses on MP3-weblogs. Out of the crawling process, each feed item is represented as a text document, containing the content of the item, as well as the links to the audio files. Then, classic text retrieval system outputs relevant feed items related to the user's query. Furthermore, a content-based navigation allows users to browse through the retrieved items and discover new music and artists using audio similarity.

Ongoing work includes the automatic extraction of music related tags (i.e. *guitar*, *rock*, 1970s) from the text, as well as applying autotagging to incoming audio files; using audio content-based similarity [6]. We also plan to add relevance feedback to tune the system and get more accurate results, specially for the content-based similarity.

The system is available at http://www.searchsounds.net.

8.2 FOAFing the Music: Music Recommendation in the Long Tail

Now we present the second of the two prototypes developed. It is a music recommender system, named *FOAFing the Music*, that allows users to discover a wide range of music located along the Long Tail. The system exploits music related information that is being syndicated (as RSS feeds) on thousands of websites. Using the crawled information, the system is able to filter it and recommend it to the user, according to her profile and listening habits.

8.2.1 Motivation

The World Wide Web has become the host and distribution channel for a broad variety of digital multimedia assets. Although the Internet infrastructure allows simple straightforward acquisition, the value of these resources lacks powerful content management, retrieval and visualisation tools. Music content is no exception: although there is a sizeable amount of text-based information related to music (album reviews, artist biographies, etc.) this information is hardly ever associated with the objects it refers to, that being the music files themselves (MIDI or audio). Moreover, music is an important vehicle for communicating to other people something relevant about our personality, history, etc.

There is a clear interest in the Semantic Web field in creating a Web of machine-readable homepages describing people, the links among them, and the things they create and do. The Friend of a Friend (*Friend Of A Friend*) project[3] provides conventions and a language to describe homepage-like content and social networks. The Friend of a Friend vocabulary provides properties and classes for describing common features of people and their social networks. Friend of a Friend is based on the Resource Description Framework (RDF[4]) vocabulary.

We foresee that with a complete user's Friend of a Friend profile, our system would get a better representation of the user's musical needs. On the other hand, the RSS vocabulary[5] allows systems one to syndicate Web content on the Internet. Syndicated content includes data such as news, event listings, headlines, project updates, as well as music related information, such as new music releases, album reviews, podcast sessions, and upcoming gigs.

To our knowledge, nowadays it does not exist any system that recommends items to a user, based on her Friend of a Friend profile. Yet, it is worth to mention the *FilmTrust* system.[6] It is a part of a research study aimed to understanding how social preferences might help web sites to present information in a more useful way [7]. The system collects user reviews and ratings about movies, and holds them into the user's Friend of a Friend profile [8].

8.2.2 Goals

The main goal of the *FOAFing the Music* system is to recommend, to discover and to explore music content; based on user profiling (via Friend of a Friend descriptions), context based information (extracted from music related RSS feeds), and content based descriptions (automatically extracted from the audio itself). All of that being based on a common ontology that describes the musical domain.

Figure 8.4 shows the relationship between the music information plane, and the different sources of metadata that the system exploits. Compared to the first prototype (*Searchsounds*), *Foafing the Music* holds a user profile representation, based on the Friend of a Friend initiative (already presented in Sec. 3.2). A Friend of a Friend user profile allows to filter music related information according to user's preferences.

---

[3] http://www.foaf-project.org

[4] http://www.w3.org/RDF

[5] http://web.resource.org/rss/1.0/

[6] http://trust.mindswap.org/FilmTrust

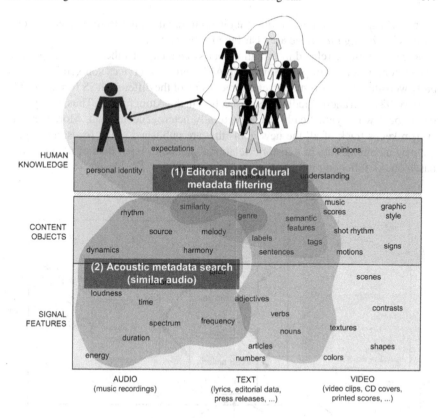

Fig. 8.4 *FOAFing the Music* and the music information plane.

8.2.3 System Overview

The overview of the *Foafing the Music* system is depicted in Fig. 8.5. The system is divided in two main components, that is (i) how to gather data from external third party sources (presented in Sec. 8.2.3.1), and (ii) how to recommend music to the user based on the crawled data, and the semantic description of the music titles (Sec. 8.2.3.3).

8.2.3.1 Gathering Music Related Information

Personalised services can raise privacy concerns due to the acquisition, storage and application of sensitive personal information [9]. In our system, information about the user is not stored in the system in any way. Instead, the system has only a link pointing to the user's Friend of a Friend profile (often a link to a *Livejournal* account). Thus, the sensitivity of this data is up to the user, not to the system. Users' profiles in *Foafing the Music* are distributed over the net.

Regarding music related information, our system exploits the mashup approach. The system uses a set of public available APIs and web services sourced from third party websites. This information can come in any of the different RSS formats (v2.0, v1.0, v0.92 and Yahoo! Media RSS), as well as in the Atom format. Thus, the system has to deal with syntactically and structurally heterogeneous data. Moreover, the system keeps track of all the new items that are published in the feeds, and stores the new incoming data in a historic relational database. Input data of the system is based on the following information sources:

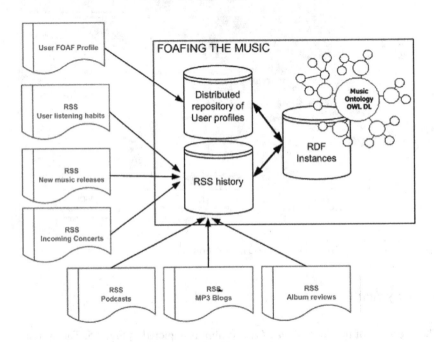

Fig. 8.5 Architecture of the *Foafing the Music* system.

- *User listening habits.* To keep track of the user's listening habits, the system uses the services provided by *last.fm*. This system offers a list of RSS feeds that provide the most recent tracks a user has played. Each item feed includes the artist name, the song title, and a timestamp—indicating when the user has listened to the track.
- *New music releases.* The system uses a set of RSS feeds that gathers new music releases from *iTunes, Amazon, Yahoo! Shopping* and *Rhapsody*.
- *Upcoming concerts.* The system uses a set of RSS feeds that syndicates music related events. The websites are: *Eventful.com*, and *Upcoming.org*. Once the system has gathered the new items, it queries the Google Maps API to get the

geographic location of the venues, so it can be filtered according to the user's location.

- *Podcast sessions.* The system gathers information from a list of RSS feeds that publish podcast sessions.
- *MP3 Blogs.* The system gathers information from a list of MP3 blogs that talk about artists and new music releases.
- *Album reviews.* Information about album reviews are crawled from the RSS feeds published by *Rateyourmusic.com, Pitchforkmedia.com,* online magazines *Rolling Stone,*[7] *BBC,*[8] *New York Times,*[9] and *75 or less records.*[10]

Table 8.1 shows some basic statistics of the data that has been gathered since mid April, 2005 until the first week of March, 2010. These numbers show that the system has to deal with daily incoming data.

Source	# RSS seed feeds	# Items stored
New releases	44	1,283,640
MP3 blogs	127	991,997
Podcasts	833	288,992
Album reviews	18	206,265
Upcoming concerts	16	369,651

Table 8.1 Information gathered from music related RSS feeds is stored into a relational database. Based on the user's Friend of a Friend profile, the system filters this information, and presents the most relevant items according to her musical taste.

8.2.3.2 Music Ontologies

An ontology is an explicit and formal specification of a conceptualisation [10]. In general, an ontology describes formally a domain of discourse. The requeriments for Ontology languages are: a well-defined syntax, a formal semantics, and a reasoning support that checks the consistency of the ontology, checks for unintended relationships between classes, and automatically classifies instances in classes.

The Web Ontology Language (OWL[11]) has a richer vocabulary description language for describing properties and classes than RDF Schema (RDFS[12]). OWL has relations between classes, cardinality, equality, characteristics of properties and enumerated classes. The OWL language is build on top of RDF and RDFS, and uses RDF/XML syntax. OWL documents are, then, RDF documents.

---

[7] http://www.rollingstone.com/

[8] http://www.bbc.co.uk/

[9] http://www.nytimes.com/

[10] http://www.75orless.com/

[11] http://www.w3.org/TR/owl-guide/

[12] http://www.w3.org/TR/rdf-schema/

On the other hand, we have defined a simple music recommendation OWL DL ontology[13] that describes some basic properties of the artists and music titles, as well as some descriptors automatically extracted from the audio files (e.g. tonality, rhythm, moods, music intensity, etc.). In [11] we propose a way to map our ontology and the *Musicbrainz* ontology, onto the MPEG-7 standard, which acts as an upper-ontology for multimedia description. This way we can link our dataset with the *Musicbrainz* information in a straightforward manner.

A focused web crawler has been implemented to add instances to our music ontology. The crawler extracts metadata of artists and songs, and the relationships between artists (such as: "related with", "influenced by", "followers of", etc.), and converts it to RDF/XML notation. The seed sites to start the crawling process are music metadata providers, such as *MP3.com, Yahoo! Music*, and *RockDetector*, as well as independent music labels (*Magnatune, CDBaby, Garageband*, etc.).

Based on our lightweight music recommendation ontology, listing 8.2 shows the RDF/XML description of an artist from *GarageBand*.

```
<rdf:Description rdf:about="http://www.garageband.com/artist/
    randycoleman">
 <rdf:type rdf:resource="{\&}music;Artist"/>
 <foaf:name>Randy Coleman</foaf:name>
 <music:decade>1990</music:decade>
 <music:decade>2000</music:decade>
 <music:genre>Pop</music:genre>
 <foaf:based_near
    rdf:resource="http://sws.geonames.org/5368361/"/>
 <music:influencedBy
    rdf:resource="http://www.coldplay.com"/>
 <music:influencedBy
    rdf:resource="http://www.jeffbuckley.com"/>
 <music:influencedBy
    rdf:resource="http://www.radiohead.com"/>
</rdf:Description>
```

Listing 8.2 RDF example of an artist individual

Listing 8.3 shows the description of an individual track of the previous artist, including basic editorial metadata, and some features extracted automatically from the audio file.

```
<rdf:Description rdf:about="http://www.garageband.com/song?|pe1|
    S8LTM0LdsaSkaFeyYG0">
 <rdf:type rdf:resource="{\&}music;Track"/>
 <music:title>Last Salutation</music:title>
 <music:playedBy rd:resource="http://www.garageband.com/artist/
    randycoleman"/>
 <music:duration>247</music:duration>
 <music:intensity>Energetic</music:intensity>
 <music:key>D</music:key>
 <music:keyMode>Major</music:keyMode>
 <music:tonalness>0.84</music:tonalness>
 <music:tempo>72</music:tempo>
</rdf:Description>
```

---

[13] http://foafing-the-music.iua.upf.edu/music-ontology#

Listing 8.3 Example of a track individual

These individuals are used in the recommendation process, to retrieve artists and songs related with the user's musical taste.

8.2.3.3 Providing Music Recommendations

This section explains the music recommendation process, based on all the information that has continuously been gathered from the RSS feeds and the crawler. Music recommendations, in the *Foafing the Music* system, are generated according to the following steps:

1. Get music related information from user's Friend of a Friend interests, and listening habits from *last.fm*,
2. Detect artists and bands,
3. Compute similar artists, and
4. Rate the results by relevance, according to the user's profile.

To gather music related information from a Friend of a Friend profile, the system extracts the information from the FOAF interest property (if dc:title is given then it gets its value, otherwise it gathers the text from the <title> tag of the HTML resource).

```
<foaf:interest
   rdf:resource="http://www.tylaandthedogsdamour.com/"
   dc:title="The_Dogs_d'Amour" />
```

Listing 8.4 Example of a Friend of a Friend interest with a given dc:title.

The system can also extract information from a user's Friend of a Friend interest that includes the artist description based on the general Music Ontology [12].

The following example presents a way to express interest in an artist, by means of the general Music Ontology.

```
<foaf:interest>
   <mo:MusicArtist rdf:about='http://musicbrainz.org/artist/12
      d432a3-...-d20751880764'>
      <mo:discogs rdf:resource='http://www.discogs.com/artist/Yann+
         Tiersen'/>
      <foaf:img rdf:resource='http://ec2.images-amazon.com/images/P
         /B000852GIQ...Z_.jpg'/>
      <foaf:homepage rdf:resource='http://www.yanntiersen.com/'/>
      <foaf:name>Yann Tiersen</foaf:name>
      <mo:wikipedia rdf:resource='http://en.wikipedia.org/wiki/
         Yann_Tiersen'/>
   </mo:MusicArtist>
</foaf:interest>
```

Listing 8.5 FOAF example of an artist description that a user is interested in.

Based on the music related information gathered from the user's profile and listening habits, the system detects the artists and bands that the user is interested in, by doing a SPARQL query to the artist RDF repository. Once the user's artists have been detected, artist similarity is computed. This process is achieved by exploiting the RDF graph of artists' relationships (e.g. *influenced by*, *followers of*, *worked with*, etc.), as shown in Listing 8.2.

The system offers two ways of recommending music information. On the one hand, *static* recommendations are based on the favourite artists encountered in the Friend of a Friend profile. We assume that a Friend of a Friend profile would be rarely manually updated or modified. On the other hand, *dynamic* recommendations are based on user's listening habits, which are updated much more often than the user's profile. Following this approach a user can discover a wide range of new music and artists on a daily basis.

Once the recommended artists have been computed, *Foafing the Music* filters music related information coming from the gathered music information (see Sec. 8.2.3.1) to:

- Get new music releases from iTunes, Amazon, Yahoo Shopping, etc.
- Download (or stream) audio from MP3-blogs and Podcast sessions,
- Create, automatically, XSPF[14] playlists based on audio similarity,
- View upcoming gigs happening near to the user's location, and
- Read album reviews.

Syndication of the website content is done via an RSS 1.0 feed. For most of previous functionalities, there is a feed subscription option to get the results.

8.2.3.4 Usage Data

Since its inception in August 2005, the system has an average of 60 daily unique accesses, from more than 5,000 registered users, including casual users that try the *demo* option. More than half of the users automatically created an account using an external Friend of a Friend profile (most of the times, around 70%, the profile came from their *Livejournal* Friend of a Friend account). Also, more than 65% of the users add her *last.fm* account, so we can use their listening habits from *last.fm*. Figure 8.6 shows the number of logins over time, since August 2005 till July 2008. The peaks are clearly correlated with related news about the project (e.g. local TV and radio interviews, and reviews on the web).

8.2.4 Summary

We have proposed a system that filters music related information, based on a given user's Friend of a Friend profile and her listening habits. A system based on Friend of a Friend profiles and user's listening habits allows the system to "understand" a

---

[14] http://www.xspf.org/. XSPF is a playlist format based on XML syntax

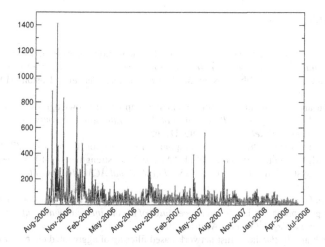

Fig. 8.6 Daily accesses to *Foafing the Music*. The system has an average of 60 daily unique accesses, from more than 4,000 registered users and also casual users that try the *demo* option.

user in two complementary ways; psychological factors—personality, demographic preferences, social relationships—and explicit musical preferences. In the music field, we expect that filtering information about new music releases, artists' interviews, album reviews, and so on, can improve user satisfaction as it provides the context and needed information to backup the system's recommendations.

Describing music assets is a crucial task for a music recommender system. The *success* of a music recommender can depend on the accuracy and level of detail of the musical objects, and its links within a user profile. Furthermore, we formalise into an ontology the basic musical concepts involved in the recommendation process. Linking these musical objects with the user profile eases the recommendation process.

Furthermore, high–level musical descriptors can increase the accuracy of content retrieval, as well as provide better personalised recommendations. Thus, going one step beyond, it would be desirable to combine mid–level acoustic features with as much editorial and cultural metadata as possible. From this combination, more sophisticated inferences and semantic rules would be possible. These rules could derive hidden high–level metadata that could be easily understood by the end-user, also enhancing their profiles. Since the existence of the general Music Ontology (MO) [12], we foresee that linking our recommendation ontology with it, as well as using all the linked information available in the Web of Data,[15] we can improve our recommender, becoming a truly semantically-enhanced music recommender.

Foafing the Music is available at http://foafing-the-music.iua.upf.edu.

---

[15] See http://linkeddata.org/.

References

1. I. Knopke, "Aroooga: An audio search engine for the world wide web," in *Proceedings of 5th International Conference on Music Information Retrieval*, (Barcelona, Spain), 2004.
2. R. Baeza-Yates and B. Ribeiro-Neto, *Modern Information Retrieval*. Boston, MA: Addison-Wesley, 1st edn., 1999.
3. S. Vembu and S. Baumann, "A self-organizing map based knowledge discovery for music recommendation systems," in *Proceedings of the 2nd International Symposium on Computer Music Modeling and Retrieval*, (Esbjerg, Denmark), 2004.
4. M. F. Porter, "An algorithm for suffix stripping," *Program,* vol. 14, pp. 130–137, 1980.
5. P. Cano, M. Koppenberger, and N. Wack, "An industrial-strength content-based music recommendation system," in *Proceedings of 28th International ACM SIGIR Conference*, (Salvador, Brazil), 2005.
6. M. Sordo, C. Laurier, and O. Celma, "Annotating music collections how content-based similarity helps to propagate labels," in *Proceedings of the 8th International Conference on Music Information Retrieval*, (Vienna, Austria), 2007.
7. J. Golbeck and B. Parsia, "Trust network-based filtering of aggregated claims," *International Journal of Metadata, Semantics and Ontologies*, vol. 1, no. 1, 2005.
8. J. Golbeck, *Computing and Applying Trust in Web-based Social Networks*. PhD thesis, College Park, MD, 2005.
9. E. Perik, B. de Ruyter, P. Markopoulos, and B. Eggen, "The sensitivities of user profile information in music recommender systems," in *Proceedings of Private, Security, Trust*, 2004.
10. T. R. Gruber, "Towards principles for the design of ontologies used for knowledge sharing," in *Formal Ontology in Conceptual Analysis and Knowledge Representation* (N. Guarino and R. Poli, eds.). Deventer, The Netherlands: Kluwer Academic Publishers, 1993.
11. R. Garcia and O. Celma, "Semantic integration and retrieval of multimedia metadata," in *Proceedings of 4th International Semantic Web Conference. Knowledge Markup and Semantic Annotation Workshop*, (Galway, Ireland), 2005.
12. Y. Raimond, S. A. Abdallah, M. Sandler, and F. Giasson, "The music ontology," in *Proceedings of the 8th International Conference on Music Information Retrieval*, (Vienna, Austria), 2007.

Chapter 9
Conclusions and Further Research

Research in recommender systems is multidisciplinary. It includes several areas, such as: search and filtering, data mining, personalisation, social networks, text processing, complex networks, user interaction, information visualisation, signal processing, and domain specific models, among others. Furthermore, current research in recommender systems has strong industry impact, resulting in many practical applications.

We have an overwhelming number of choices about which music to listen to. As stated in *The Paradox of Choice* [1], we—as consumers—often become paralysed and doubtful when facing the overwhelming number of choices. The main problem is the awareness of content, not the actual access to the content (think on *Spotify* or *YouTube* services, for example). Personalised filters and recommender systems are key elements in this scenario. Effective recommendation systems should promote novel and relevant material (non-obvious recommendations), taken primarily from the tail of the music popularity distribution.

One of the main goals in this book is music discovery via the functionality that recommender systems offer. In this sense, novelty and relevance of the recommendations are the two most important aspects. We make use of the Long Tail shape to model the popularity bias that exists in any recommender system, and use this data to recommend unknown items, hidden in the tail of the popularity curve. Our experience is that using the $F(x)$ function (see Chap. 4) to model the Long Tail curve, we get more accurate results than fitting the curve to well-known distributions, such as a power-law or log-normal [2].

Music is somewhat different from other entertainment domains, such as movies or books. Tracking users' preferences is mostly done implicitly, via their listening habits (instead of asking users to explicitly rate the items). Any user can consume an item (e.g., a track or a playlist) several times, even repeatedly and continuously. Regarding the evaluation process, music recommendation allows users instant feedback via brief audio excerpts. Thus, we have proposed new approaches to evaluate the effectiveness of the recommendations in the music domain. The evaluation focuses on the central pillar of any recommender system: the similarity among objects (e.g. items or users). In our case, we evaluate and analyse the artist and user

Ò. Celma, *Music Recommendation and Discovery*,
DOI 10.1007/978-3-642-13287-2_9, © Springer-Verlag Berlin Heidelberg 2010

similarity networks. We also present a survey with 288 subjects that provided feedback about the (personalised) recommendations. This survey evaluates the user's perceived quality and novelty factor of the music recommended.

9.1 Book Summary

This book presents a number of novel ideas that address existing limitations in recommender systems, and the lack of systematic methods to evaluate the novelty and perceived quality of recommendations in the music domain. Furthermore, two real web-based systems have been implemented to demonstrate the ideas derived from the theoretical work. The main outcomes of the book are:

1. A novel network-centric evaluation method for recommender systems, based on the analysis of the item (or user) similarity graph, and the combination with items' popularity, using the Long Tail curve. An exhaustive study comparing different approaches of music recommendation networks is presented in Chap. 6.
2. A user-centric evaluation, based on the immediate feedback of the provided recommendations, that measures the user's perceived quality and novelty factor of the recommendations. An in-depth user-based evaluation of three different music recommendation approaches is presented in Chap. 7.
3. A music search engine, named *Searchsounds*, that allows users to discover unknown music that is available on music related blogs.
4. A system prototype, named *FOAFing the music*, that provides music recommendation based on the user preferences and listening habits.

The first two contributions are more scientific, whilst the third and fourth are more engineering oriented.

9.1.1 Scientific Contributions

9.1.1.1 A Network-Based Evaluation Method for Recommender Systems

We have formulated a network-based evaluation method for recommender systems, based on the analysis of the item (or user) similarity network, combined with item popularity. This method has the following advantages:

1. It measures the novelty component of a recommendation algorithm.
2. It models the item popularity curve.
3. It combines both the complex network and the item popularity analysis to determine the underlying characteristics of the recommendation algorithm.
4. It does not require any user intervention in the evaluation process.

We have applied the network-based analysis to two different similarity graphs; for artists, and users. The results from the artist network analysis show that the *last.fm* social-based recommender tends to reinforce popular artists, at the expense of discarding less-known music. Thus, the popularity effect derived from the community of users has consequences in the recommendation network. This reveals a somewhat poor discovery ratio when just browsing through the network of similar music artists. *Allmusic.com* expert-based recommendations are more expensive to create, and also have a smaller Long Tail coverage, compared to automatically generated recommendations like collaborative filtering or audio content-based similarity. Regarding popularity, the hubs in the expert network are comprised of mainstream music. Our guess is that the editors connect long tail artists with the most popular ones, either for being influential or because many bands are considered *followers of* these mainstream artists. An audio content-based similarity network is not affected by the popularity bias of the artists, however it is prone to the musical genre biases of the collection, where the predominant genres includes most of the similar artists. The main problem of audio content-based systems is the assumption that just because two songs sound similar, any user will like both. It is very unlikely that a user will love both a *Franz Schubert*'s piano sonata, and a *Meat Loaf* piano ballad (such as "Heaven Can Wait") just because the two contain a prominent piano melody.

The results from the user network analysis show that user similarity network derived from collaborative filtering resembles a social network, whilst the network derived from audio content-based similarity has the properties of a hierarchy, where a few nodes connect small clusters. The authorities in the CB network are the users that listen to more music, independently of the quality or popularity of the music they listen to. Contrastingly, the authorities in the CF network are the users that listen to more mainstream music. These considerations have a big impact on recommendation algorithms that compute recommendations by means of user neighbourhood information.

9.1.1.2 A User-Based Evaluation Method for Recommender Systems

Our proposed evaluation measures the user's perceived quality and novelty of the recommendations. The user-centric evaluation approach has the following advantages:

1. It measures the novelty factor of a recommendation algorithm considering the user's knowledge of the items.
2. It measures the perceived quality (e.g., like it or not) of the recommendations.
3. Users provide immediate feedback to the evaluation system, so the algorithm can adapt accordingly.

This method complements the previous, user-agnostic, network-based evaluation approach. We use the user-centric method to evaluate and compare three different music recommendation approaches. In this experiment, 288 subjects rated the recommendations in terms of novelty (*does the user know the recommended song/artist?*), and relevance (*does the user like the recommended song?*).

The results from the music recommendation survey show that, in general, users' perceived quality for novel recommendations is neutral or negative (mean rating around 3/5 or less). This emphasises the need for adding context when recommending unknown music. Recommender systems should give as many reasons as possible to support their decisions.

In terms of algorithms, the rating scores for the *last.fm* social-based approach are higher than those for the hybrid and pure audio content-based similarity. However, the social-based recommender generates more familiar (less novel) songs than CB and HY. Thus, content-based and hybrid approaches provide more novel recommendations, although their quality is not as good as the ones from *last.fm*.

9.1.2 Industrial Contributions

9.1.2.1 *FOAFing the Music*: A Music Recommendation System

The system prototype, named *FOAFing the Music*, provides music recommendation based on the user preferences and listening habits. The main goal of *FOAFing the Music* is to recommend, to discover and to explore music content via user profiling, context-based information (extracted from music related RSS feeds), and content-based descriptions (automatically extracted from the audio itself). The system has an average of 60 daily unique accesses, from more than 5,000 registered users and also casual users that try the *demo* option. *FOAFing the music* allows users to:

1. get new music releases from *iTunes, Amazon, Yahoo Shopping*, etc.
2. download (or stream) audio from MP3-blogs and Podcast sessions,
3. discover music with *radio-a-la-carte* (i.e. personalised playlists),
4. view upcoming nearby concerts, and
5. read album reviews.

Since the existence of the general Music Ontology [3], we foresee that linking our recommendation ontology with it, as well as exploiting all the linked information available in the Web of Data,[1] we can improve our system, becoming a truly semantically-enhanced music recommender.

9.1.2.2 *Searchsounds*: A Music Search Engine

We have implemented a music search engine, named *Searchsounds*, that allows users to discover unknown music mentioned on music-related blogs. *Searchsounds* provides keyword based search, as well as the exploration of similar songs using

---

[1] See http://linkeddata.org/.

audio similarity. The system allows users to dig into the Long Tail, by providing music discovery using audio content-based similarity, that could not be easily retrieved using classic text retrieval techniques. Over 400,000 audio files are currently indexed, using both text and audio features.

Ongoing work includes the automatic extraction of music related tags (i.e. *guitar*, *rock*, 1970s) from the text, as well as applying autotagging to incoming audio files; using audio content-based similarity [4].

9.2 Limitations and Further Research

9.2.1 Dynamic Versus Static Data

It goes without saying that there are many ways in which the work presented in this book could be extended or improved. One of the main limitations of our approach is that it is not *dynamic*. We work with a *snapshot* of the item (or user) similarity network, and the analysis is based on this data. However, the recommendation network dynamics is an important aspect of any recommender system. Users' taste change over time, and so it does the similarity among items. Further work in this area would include a detailed study of a dynamic model in the network—including trend and hype-item detection—and a comparison with our stationary model.

9.2.2 Domain Specific

The work done has been applied only to music recommendation. Even though we did not use any domain-specific metrics in the network-centric evaluation, our findings cannot be directly extrapolated to other domains. Further work could be to extend the network-centric experiments to other domains, such as movie recommendation using the *Netflix* dataset.

Besides, the user-centric evaluation contains a lot of particularities from the music recommendation domain. In other domains (e.g., movies, books, or travels), explicit user feedback about the recommended items cannot be provided in real-time. Furthermore, our music recommendation survey design is based on providing blind recommendations. Future work should be to compare our results with a new experiment that provides contextual information and transparency about the music being recommended. The related question would be whether the ratings of novel items increase (i.e. are perceived with better quality) when providing more information about the recommended songs.

9.2.3 User Evaluation

In our user-centric evaluation we could not classify a participant into the four type of listeners (savant, enthusiasts, casuals and indifferents). In fact, it would be interesting to look at recommendation evaluations through the lense of the four types of listeners. The type and utility of recommendations varies greatly depending on the type of user. When testing against the general population—since most listeners fall into the casual or indifferent bucket—recommenders that appeal to these types of listeners would score well when compared to recommenders that are designed for the enthusiast or savant. However, enthusiasts and savants are likely to be much more active consumers, so from an economic point of view, there may be more value targeting them. Recommenders for savants and enthusiasts would probably favour novelty and long tail content, while recommendations for a casual listener would probably favour low-risk exploration. Indeed, a new task for music recommenders could be to help casual listeners appreciate diversity and exploration to unknown content.

9.2.4 User Understanding

User understanding is another important aspect when providing personalised recommendations. Our approach to model a user profile is a rather simple list of preferred artists. Extending the user profile model, adding relevant and contextual information, would allow recommender systems to have a better understanding of the user.

Ideally, a recommender system should provide different and personalised recommendations for a given item. That is, when visiting the *Beatles' White Album* in Amazon store, the system should present the list of recommendations according to the user profile. Depending on the user's taste, the system should stress the pop side of the band, whilst in other situations it could promote the more psychedelic or experimental music they did. Ongoing work by Lamere and Maillet [5] is aligned with this idea. They have implemented a prototype system that creates transparent, steerable recommendations. Users can modify the list of recommended artists, by changing the tag cloud of the seed artist. This way, users focus on some particular styles or aspects of the musician (e.g. give me *The Beatles* similar artists, but emphasizing their psychedelic rock side).

9.2.5 Recommendations with No Explanation

Blind recommendations do not provide any context nor explanation. Thus, it does not help in assessing the relevance of novel recommendations. It might be the case that some novel songs recommended are perceived as non-relevant, but when explaining the ties with the user profile the perceived quality could be increased. In fact, *why* is as important as *what* is being recommended. Again, [5] is a novel example of a system that gives transparent explanations about the provided recommendations.

9.3 Outlook

We are witnessing an explosion of practical applications coming from Music Information Retrieval research field: music identification systems, music recommenders and playlist generators, music search engines, extraction of semantic audio features, autotagging, and this is just the beginning.

A few years ago, music was a key factor in taking the Internet from its text-centered origins to being a complete multimedia environment. Music might do the same for the next web generation. The "Celestial Jukebox" is about to become a reality.

References

1. B. Schwartz, *The Paradox of Choice: Why More Is Less*. New York, NY: Harper Perennial, January 2005.
2. K. Kilkki, "A practical model for analyzing long tails," *First Monday*, vol. 12, May 2007.
3. Y. Raimond, S. A. Abdallah, M. Sandler, and F. Giasson, "The music ontology," in *Proceedings of the 8th International Conference on Music Information Retrieval*, (Vienna, Austria), 2007.
4. M. Sordo, C. Laurier, and O. Celma, "Annotating music collections how content-based similarity helps to propagate labels," in *Proceedings of the 8th International Conference on Music Information Retrieval*, (Vienna, Austria), 2007.
5. P. Lamere and F. Maillet, "Creating transparent, steerable recommendations," in *Late-breaking Demos. Proceedings of the 8th International Conference on Music Information Retrieval*, (Philadelphia, PA), 2008.

Index